Make:
EDUCATION
FORUM
A VIRTUAL EVENT BY AND FOR MAKER EDUCATORS
During this innovative two-day virtual event, you'll dive into exciting maker topics, connect with like-minded educators, and explore new technical skills that will transform how you teach.
SEPTEMBER 22—23, 2023
make.co/educationforum
T0093734

Make: 85

CONTENTS

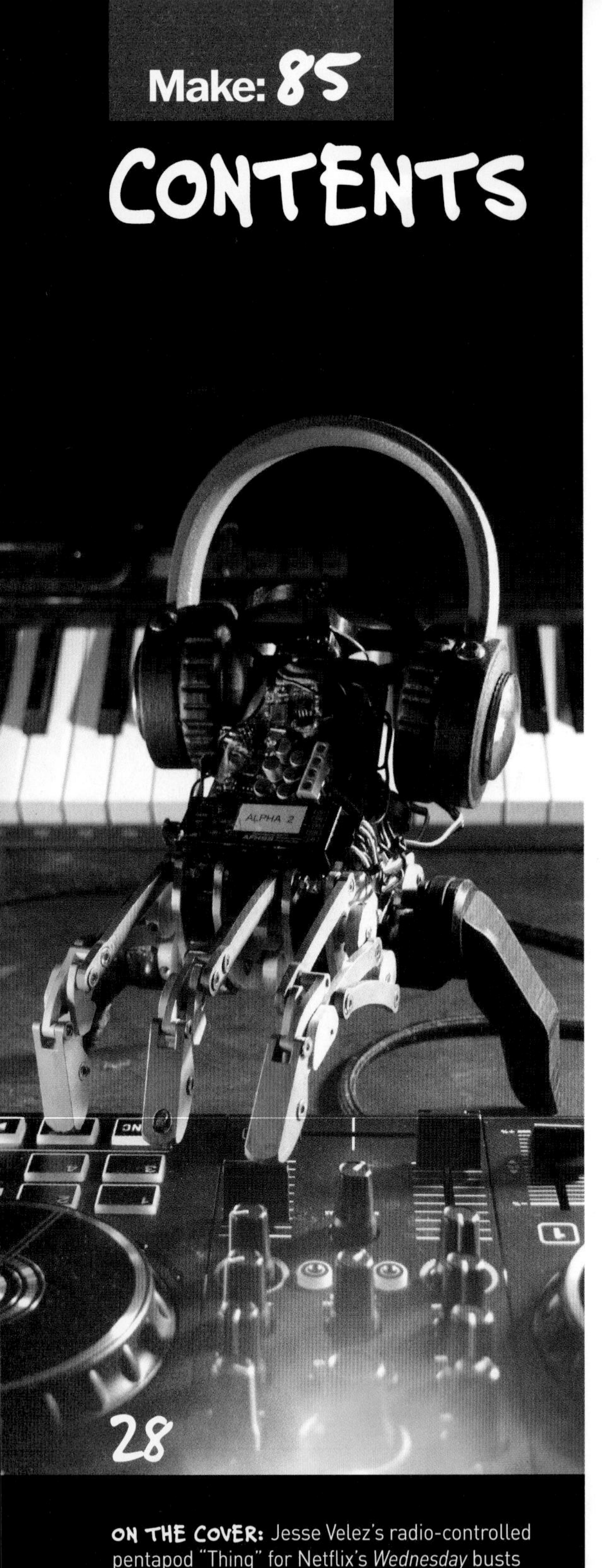

28

ON THE COVER: Jesse Velez's radio-controlled pentapod "Thing" for Netflix's *Wednesday* busts out some beats.

Photos: Jesse Velez and Charlyn Gonda

COLUMNS

FEATURES

RADIO CONTROL

DIY MUSIC

PROJECTS

SKILL BUILDER

TOOLBOX

OVER THE TOP

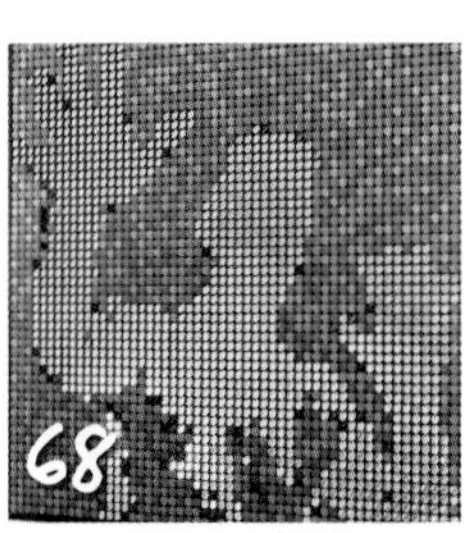

Jesse Valez, Kelly Heaton, Victor Chaney, Charles Saadiq, Owen McAteer, Sean Nolan, Rich Cameron

Make:®

"Musical instruments provide the most efficient and refined interface between men and machine of anything we know." —*Robert Moog*

PRESIDENT
Dale Dougherty
dale@make.co

VP, PARTNERSHIPS
Todd Sotkiewicz
todd@make.co

EDITORIAL

EDITOR-IN-CHIEF
Keith Hammond
keith@make.co

SENIOR EDITOR
Caleb Kraft
caleb@make.co

COMMUNITY EDITOR
David J. Groom
david@make.co

PRODUCTION MANAGER
Craig Couden

CONTRIBUTING EDITORS
Tim Deagan
William Gurstelle

CONTRIBUTING WRITERS
Rhett Allain, Cabe Atwell, Joe Bauer, Rich Cameron, Victor Chaney, Jet Kye Chong, Ben Eadie, Nick Gaydos, Greg Gilman, Charlyn Gonda, Sean Hallowell, Brad Halsey, Kelly Heaton, John Ivener, Bob Knetzger, Andy Lee, Bill Van Loo, Owen McAteer, Forrest M. Mims III, Sean Nolan, Kirk Pearson, Marshall Piros, Charles Platt, Michael Seltzer, Wayne Seltzer, Becky Stern, Jesse Velez, Yuri Vlasyuk, Lee Wilkins, Lee D. Zlotoff

CONTRIBUTING ARTISTS
Jesse Velez

MAKE.CO

ENGINEERING MANAGER
Alicia Williams

WEB APPLICATION DEVELOPER
Rio Roth-Barreiro

DESIGN

CREATIVE DIRECTOR
Juliann Brown

BOOKS

BOOKS EDITOR
Kevin Toyama
books@make.co

GLOBAL MAKER FAIRE

MANAGING DIRECTOR, GLOBAL MAKER FAIRE
Katie D. Kunde

GLOBAL LICENSING
Jennifer Blakeslee

MARKETING

DIRECTOR OF MARKETING
Gillian Mutti

PROGRAM COORDINATOR
Jamie Agius

OPERATIONS

ADMINISTRATIVE MANAGER
Cathy Shanahan

ACCOUNTING MANAGER
Kelly Marshall

OPERATIONS MANAGER & MAKER SHED
Rob Bullington

LOGISTICS COORDINATOR
Phil Muelrath

PUBLISHED BY

MAKE COMMUNITY, LLC
Dale Dougherty

Printed in the U.S. by Schumann Printers, Inc.

Comments may be sent to:
editor@makezine.com

Visit us online:
make.co

Follow us:
@make @makerfaire
@makershed
makemagazine
makemagazine
makemagazine
makemagazine

Manage your account online, including change of address: makezine.com/account
For telephone service call 847-559-7395 between the hours of 8am and 4:30pm CST.
Fax: 847-564-9453.
Email: make@omeda.com

Make: Community

Support for the publication of *Make:* magazine is made possible in part by the members of Make: Community. Join us at make.co.

CONTRIBUTORS

What's your favorite musical artist to listen to while working in the shop?

Sean Nolan
Whidbey Island, Washington (Whittled Wonder)
I love a bit of everything (shoutout KXA 1520), but always come back to Kenny Chesney in my happy place!

Owen McAteer
Madrid, Spain (Flip-Dot Animation)
Bonobo. Sets just the right vibe and energy for me to focus and create.

Ben Eadie
Calgary, Alberta, Canada (Next Level Radio Control)
Chillhop is my jam. Honestly, I do not know any specific artists. I generally play "lo-fi" playlists on YouTube.

Issue No. 85, Summer 2023. *Make:* (ISSN 1556-2336) is published quarterly by Make Community, LLC, in the months of February, May, Aug, and Nov. Make: Community is located at 150 Todd Road, Suite 100, Santa Rosa, CA 95407. SUBSCRIPTIONS: Send all subscription requests to *Make:*, P.O. Box 566, Lincolnshire, IL 60069 or subscribe online at makezine.com/subscribe or via phone at (866) 289-8847 (U.S. and Canada); all other countries call (818) 487-2037. Subscriptions are available for $34.99 for 1 year (4 issues) in the United States; in Canada: $43.99 USD; all other countries: $49.99 USD. Periodicals Postage Paid at San Francisco, CA, and at additional mailing offices. POSTMASTER: Send address changes to *Make:*, P.O. Box 566, Lincolnshire, IL 60069. Canada Post Publications Mail Agreement Number 41129568.

FROM THE EDITOR'S DESK

READER INPUT

WHAT WAS YOUR FIRST ISSUE OF *MAKE:*?

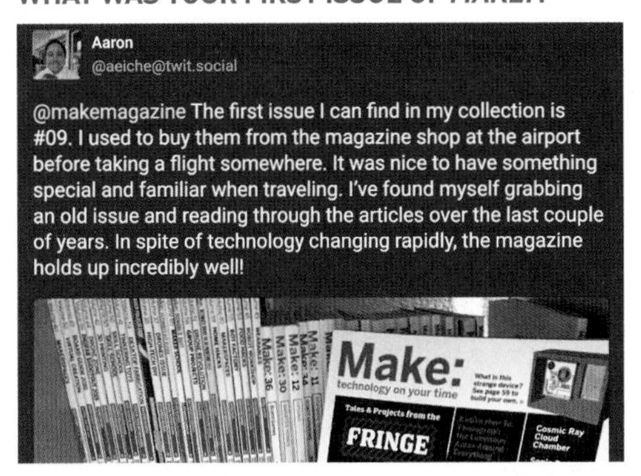

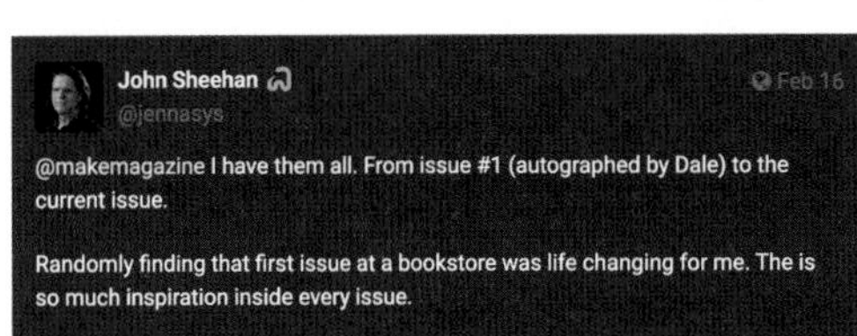

What was your first issue of *Make:* magazine? Or your favorite? Let us know at editor@makezine.com or at makezine.com/go/my-first-issue.

OH STOP IT, YOU

You have created the most fantastic issue ever! I read it cover-to-cover several times and investigated most of the generative AI apps. Also, I happen to be getting into SDR.

—Mike Winter, via email

PUTTING THE *ZINE* IN MAKEZINE

Discord user @gus shared this fun project on our Show & Tell channel. Join us on the *Make:* Discord server and show us what you're working on, at makezine.com/go/discord.

MAKE: AMENDS

In "Unleash the Amateurs" (*Make:* Volume 84, page 14), we reported that Hiram Percy Maxim contacted fellow radio amateur Windsor Locks in 1914. In fact Windsor Locks, Connecticut, is the *town* the colleague lived in. Thanks to reader Tim McNerney for spotting the error!

Seeing Both Sides Now

by Dale Dougherty, President of Make: Community

There are two sides to making, which seems to satisfy most makers.

One side is playful. Makers are enthusiasts, curious to explore, and willing to try something just for its own sake. Making is play, enjoyed as much for the process as for what it produces. Makers get to choose what projects to do, based on their interests and ideas.

The other side of making is serious, more work than play. It can be purposeful, driven by necessity or external demands. This type of making often puts the maker into a different gear and they feel a sense of accomplishment.

Yet making can be both playful and serious at the same time. A project you begin for fun becomes something you must take seriously to finish. (Just make sure you're still enjoying yourself, or the work becomes drudgery.)

Music is a great example of something fun that can also be taken very seriously. In this issue, Nick Gaydos reports how electronic music's history of experimentation has led to today's resurgence of modular synthesizers — devices you can assemble yourself that generate sound with knobs and switches. Joe Bauer shows you how to make your first synth module, a DIY Avalanche Oscillator. Bill Van Loo recommends software for making music at any level, from playful phone apps to serious digital audio workstations (DAWs). While many musicians produce their music for release, he says, "it's fine to just play for the joy of playing." Jet Kye Chong's soda bottle marimba is like that — fun to play and easy to make, an instrument made from recycled plastic that has a robust sound.

Need more synth circuits? Just for fun, Lee Wilkins hacks an Atari Punk circuit to use different fruits as resistors, and the team from Dogbotic Labs shares an amazing quirk — a simple logic chip that somehow plays notes perfectly in tune! On a more serious mission, Charles Platt shows how to generate an "unpredictable screamer circuit" to deter pesky rats, and probably your closest neighbors.

And what's more fun than a monster movie? Ben Eadie and Jesse Velez show how they create lifelike animated props for *Ghostbusters* and *Wednesday* using next-level radio control programs, with Linux running right on the transmitter.

Sometimes the skills we develop on fun projects become valuable in another context. At the outbreak of war in Ukraine in February 2022, I reconnected with Yuri Vlasyuk and Svitlana Bovkun, who had produced Maker Faires there. A year later, I asked Yuri to tell us how makers have contributed to the defense of Ukraine.

Nearly everyone in Ukraine is asked to be resourceful and help defend the country, Yuri reports. Welding is a valued skill. 3D printing is used to develop shell casings and tails for grenades. Drone racers from Maker Faire now are piloting consumer drones on the battlefield. Other makers are helping in areas where people lack shelter, heat, power, and food. The Tolocar project is a fleet of mobile makerspaces that can visit those areas to help people solve problems and learn new skills.

Brad Halsey visited Ukraine during wartime last year. He reports that, seeing their entrepreneurial energy and rapid innovation, he came back wishing America could learn to think creatively and build in the same way.

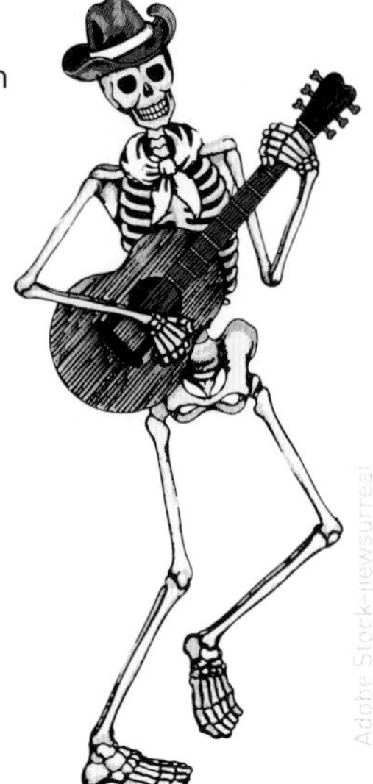

To those who perhaps unknowingly deprecate makers as hobbyists, I try to explain that making has both sides. I acknowledge that makers are proud hobbyists — and that if these hobbyists are ever called upon, they have capabilities that a society with so many problems might desperately need any day now.

CAN YOU MACGYVER A FAST FIX FOR A FLOODED FAMILY ROOM?

We are all MacGyvers now! Make: *has brought Mac back to help you think — and make — your way out of emergencies. Watch for the next challenge on our blog (makezine.com), Mastodon (@makemagazine), Twitter (@make), and Facebook (makemagazine) and enter your solutions for a chance to be featured in these pages and win* Make: *goodies!*

The Scenario

You're 17, and your family just moved into a new house in the suburbs. Your parents have an overnight date in the city, leaving you with clear instructions *not* to have a party while they're away. So, no sooner do they leave then you invite a half dozen friends over for, not a party exactly but, you know ... a *get-together*. Which, astonishingly, involves various inebriants in your finished basement cum family room. Midway through the festivities the local power goes out, driving the party upstairs to the patio outside — unaware that one of your hammered BFFs has left water running in the sink with the drain closed. When you venture back down for more refreshments at 3 a.m. you discover to your horror there's a good 2 inches of water flooding the basement! Now curiously sober, you summon your homies — who abandon you like rats from a sinking ship.

The Challenge

Now on your own, you realize you have exactly 7 hours to get the water out and clean up the mess before your folks return. And the power is still out.

What You've Got

And it's *all you've got*:

- Normal kitchen stuff. Knives, forks, spoons, cookie sheet pans, but weirdly no bowls or pots. But if it's in a kitchen drawer, you have it. Spatulas, straws, turkey baster.
- There's a nice mixer (again, no power), a toaster oven, air fryer, microwave. The fridge is off — but it has just a bare minimum: a six-pack of soda and a squeeze bottle of ketchup.
- Of course, there's the "junk drawer." You have a battery-powered flashlight, coins, keys, pocketknife, marbles, empty film canisters, AA batteries, rubber bands, paper clips.
- Yes, you have plastic straws and duct tape.
- In the garage, it's just outdoor stuff: garden hose, sprinklers, shovel, hoe, rake. There are no gas-powered devices and no buckets.
- The basement does have a small window to get out to the ground.
- The rest of the house just has normal stuff.

TURN THE PAGE FOR SOLUTIONS!

LEE ZLOTOFF is an award-winning writer, producer, and director of film and TV, including *MacGyver* (1985–1992). His new production, *MacGyver: The Musical*, casts a different audience member as Mac at each performance. macgyver.com

RHETT ALLAIN teaches physics at Southeastern Louisiana University. He was technical consultant for the *MacGyver* reboot (2016–2021) and an advisor for *MythBusters*. He blogs about physics fun at rhettallain.com.

Adobe Stock-aleutie and Reservoir Dots

MacGyver Challenge: Teenage Wasteland!

Our Solution

A siphon won't work down in this basement, so of course you need to make a pump to remove the water — but which water pump would be the best in this case? In this situation we can build a ***shake pump*** aka ***jiggle pump***, with a one-way water valve. Take the squeeze bulb off the turkey baster and drop a marble down the tube so that it sits in the narrow part. When water comes into the tube, the marble will get pushed up and let the water flow. However, if water tries to go the other way, the marble will get stuck in the narrow part of the turkey baster, trapping the water.

With the marble in the baster, you just need to connect the hose and run it outside. When you push the baster *down* into the water, it will force water *up*. When you pull the baster up, the water stays in there (because of the one-way valve). Repeat to move the water up and out! Visit youtu.be/RXdiHfsDyqo to watch how it works.

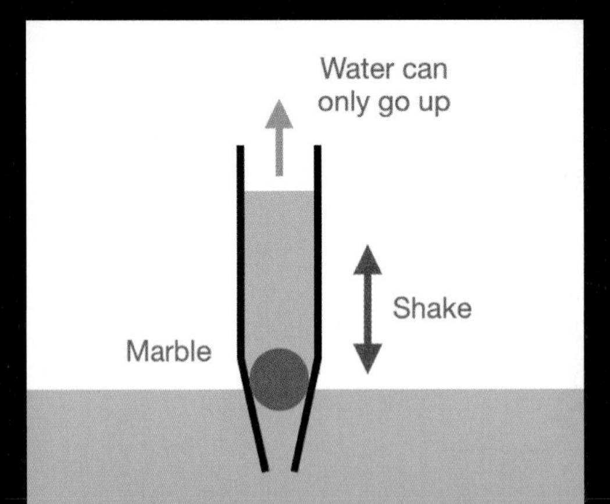

Most Plausible Solution

We are judging Chuck B.'s solution as the most plausible since it is very similar to our method and includes our turkey baster check valve. Chuck's idea includes a bit more detail, with two check valves instead of one (using the ketchup bottle too), and also with an improvised hand pump. And yes, it involves plenty of paper clips, plastic straws, and duct tape. Although this setup might take some time, it would look awesome! (Read all the full solutions at makezine.com/go/teenage-wasteland.)

Most Creative Solution

Craig Robson's idea is to make a Venturi pump using the water faucets in the house along with some hoses and parts of the ketchup bottle. The ***Venturi effect*** says that when you force a fluid through a restriction, the fluid's speed will increase — which also brings a decrease in pressure. The Venturi pump uses flowing water (from the faucet) going through a restriction (from parts of the bottle) to create suction and pull the water out of the basement.

Honorable Mention

Sometimes the best solution is no solution? David Maynor suggests that instead of removing the water, just come up with a plausible excuse for the flood: "Poke a hole in the water heater and work on your surprised face."

A LIFE WITH

Written by David J. Groom, Arm Developer Program Ambassador and Community Editor, *Make:*

Growing up in the UK, the BBC's Computer Literacy Project gave me the once-in-a-century chance to take a front-row seat in the computer revolution. At my school in Essex there was a single BBC Micro Model B in our classroom, which we were only allowed to use occasionally, and usually as a group.

The BBC Micro was the creation of Arm progenitor Acorn Computers, and its innovative "Tube" interface allowed the Acorn team to use it as a test mule while developing their next-generation RISC (Reduced Instruction Set Computer) silicon and associated software. The resultant 6MHz ARM1 (Acorn RISC Machine) chip was then dogfooded as a second processor for the BBC Micro, where it helped enable simulation, accelerate CAD work, and provide a target for a new version of BBC BASIC written in ARM assembly.

One day my father came home with Acorn's latest BBC Master 128, the successor to the venerable Model B, with which I cherished every fleeting moment. Thus began my life with Arm, though I hadn't quite realized it yet.

I recall vividly the first time I truly encountered Arm; Dad took me to a computer show where I saw, on a pedestal, the 32-bit, ARM2-powered Acorn Archimedes. The machine and several others like it were running a 3D spaceship lander (aka *Zarch*) and a real-time rendering of colored spheres to demonstrate the machine's 7x performance advantage over already impressive 68000-based contemporaries like the Amiga.

Not long after, I moved to the USA and the remainder of my formative years were spent surrounded by x86-based PCs, which never caught my passion in the same way as the Acorn devices, but saw me attending the University of Michigan Honors College to study Computer Science, and securing internships at Microsoft.

It wasn't until 2007 that I rediscovered my path to all things Arm. Falling in love with the Arduino Diecimila as part of a Roomba hacking project enamored me with embedded hardware. Rather than writing code to run on a screen or web server, it allowed me to write code that I could touch. The pinnacle of this was hacking on the Arm Cortex-M4-based Pebble smartwatch at the Pebble Rocks Boulder hackathon, which combined my passion for the unique wearable platform with my hardware obsession. The dev targets we used during the event were PJRC's Teensy boards, another Cortex-based device, meaning my team's win was thanks in part to two Arm-based chips!

With the kickoff of the **Arm Developer Program**, this year is shaping up to be a very exciting time to build on Arm. As an Arm **Ambassador** within the program, I get extensive support and insights directly from Arm experts, while the wider-reaching Arm **Developer** Program offers comprehensive resources for new and experienced developers.

Whether you're just starting out with embedded machine learning, trying to optimize your cloud computing performance and spend, or pushing the realms of mobile graphics or server performance, the Arm ecosystem is the place to be in 2023. Join me for the next ***once-in-a-century opportunity to develop the future on Arm.***

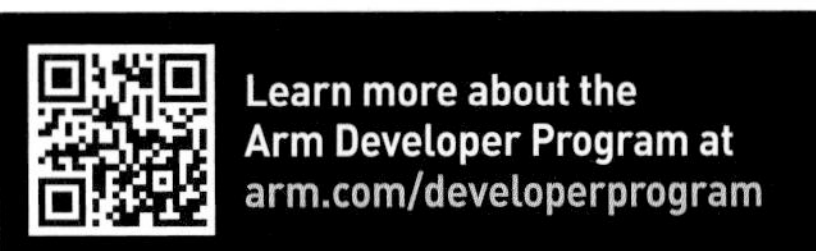

Learn more about the Arm Developer Program at arm.com/developerprogram

MADE ON EARTH

Backyard builds from around the globe

Found a project that would be perfect for Made on Earth?
Let us know: *editor@makezine.com*

BALLOONS AS ART AS THERAPY

BALLOONSINBOLD.COM

DJ Morrow is blowing up balloon art, turning a medium mostly associated with children's birthday parties into a canvas fit for fine art galleries.

The 27-year-old Texan had been twisting balloons into various characters as a Houston-based party entertainer for eight years before the Covid-19 pandemic deflated his business. Depressed and anxious about an uncertain future and dwindling finances, he reached for something he knew how to inflate, repurposing black balloons as darkness surrounding a wax hand clutching a light bulb. "It was just that idea of using art as a lifeline when you have nothing else to hold on to," he tells *Make:* of the piece *Reaching*.

It was his next piece, though, that truly convinced him balloons were capable of serious artistic expression. He recreated his favorite painting, *Saturn Devouring His Son* by Francisco Goya, and the stunning sculpture was quickly upvoted to the top of Reddit.

"I decided this is the moment where I need to just unshackle myself," Morrow says. "Stop playing towards people's expectations of what balloon art is, and just be honest and raw."

Since then, his works have only grown in scale and complexity, using a technique called balloon weaving to craft large figures, which contrast the colorful fun of the medium with darker contemplative themes of isolation and anxiety. "The new direction has given me the outlet that I needed to deal with my emotions and to have hope that things can get better — not just for me, but for society at large."

Morrow has been busy inspiring others by teaching at balloon conventions around the country while preparing for his first solo art show this spring. Time, he says, is always the biggest challenge, as balloons lose air after 30 hours, so, "once you start twisting, the clock starts running." —*Greg Gilman*

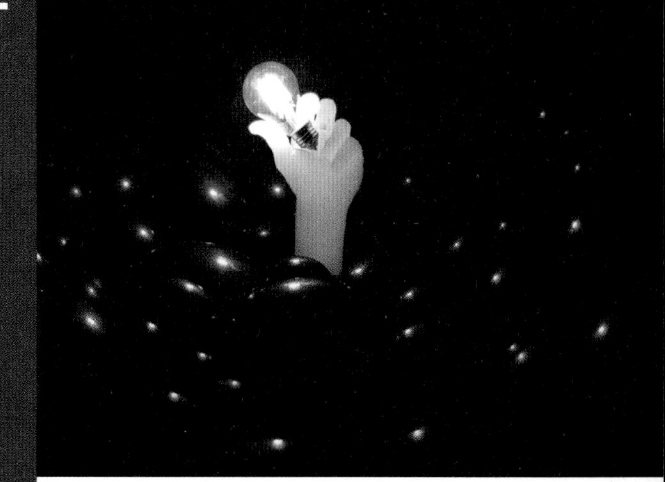

DJ Morrow

THREADING AN EYE IN THE SKY

INSTAGRAM.COM/VICTORIAROSERICHARDS

Stitching stunning sunsets and magnificent maps in miniature form, English embroidery artist **Victoria Rose Richards** takes the craft to a new level with the vibrant colors and textures in her works.

Richards, who has autism, began embroidering as a stress-relieving hobby while studying biology at the University of Exeter in her home county of Devon, England. She graduated in 2019 and still plans to someday pursue a career in that field, but she decided to continue sewing after realizing how much joy it brings her. Richards says, "It was something productive I could genuinely enjoy with no strict expectations and deadlines, no failures as such (because every failed project is just part of learning your craft!), and I could bring the pictures in my head to life."

She draws much of her inspiration from the mismatched English landscapes that are filled with ordered agrarian plots and chaotic countrysides, and often uses a top-down perspective to center a bird's-eye view of the land. Now a full-time artist for the foreseeable future, Richards' formal training in ecology and soil biology helps shape her artwork, as does her overall enthusiasm for all things nature.

As an artist whose needlework spans a variety of sizes, from 2-inch canvases to 10-inch hoops, Richards admits that she has some physical challenges making larger pieces. Besides the constant pricking of fingers, she finds that hoop stands and clamps don't suit her personal embroidery methods very well, so she has to hold her pieces up herself while she works. This places a lot of unnecessary stress on her hand and finger muscles, and she worries that too many large projects could result in long-term hand injuries.

Still, that doesn't stop Richards from remaking the English countryside in thread form, and she encourages others to do the same. "Don't be scared to experiment!" she says to those inspired by her works. "You need to go outside your comfort zone and potentially make bad art to make better art later on ... [but] first and foremost your art should make you happy to create." —*Marshall Piros*

Victoria Rose Richards

ONE LETTER AT A TIME

JAMESCOOKARTWORK.COM

Fog rolls in off the Thames River bank. It spreads through London's Trinity Buoy Wharf and its colorful shipping-container work studios. The faint sound of a typewriter fills the alleys and the streets. Its source, **James Cook**, is working late on a portrait — using a 1930s manual typewriter.

Cook is a UK-based artist who has gained worldwide recognition for his unique and intricate typewriter art. Using only the keys on one of his 63 manual typewriters, Cook creates highly-detailed works of art that range from retro-futuristic aesthetics to re-creations of seminal works from past masters.

Cook's use of different fonts and letter sizes create contrast and emphasis in pieces that often use multiple layers of typing to create texture and depth. He got his start with a secondhand 1953 Oliver Courier typewriter and prefers discontinued and antique typewriter ribbons for the color black, and sometimes red, blue, and yellow. Those with a keen eye can often find words or phrases related to the piece hidden within the art itself.

Restricted only by the typewriter carriage size, Cook is not limited to a single sheet of paper for any one work — from diptychs and triptychs to panoramic views of cityscapes, he can expand the work over multiple sheets pieced together. It takes him about a week to create a single-sheet portrait. It's a single person at a single typewriter, analog art by human hands — a rarity these days.

And a request from the artist himself: If you have a musical note typewriter, Cook would love to chat. Send him a note at instagram.com/jamescookartwork. —*Cabe Atwell*

James Cook

MAKERS IN DEFENSE OF UKRAINE

A YEAR OF FIGHTING BACK WITH INNOVATION

Learning to build a flashlight at Hacklab Kyiv

YURI VLASYUK and his wife, Svitlana Bovkun, have organized Maker Faires in the Ukraine and manage the nonprofit MakerHub.org. Yuri is chairman of the board of the Ukrainian Maker Association.

ON February 24, 2022 at about 4:50 a.m., we woke up, my wife Svitlana and I, and argued whether the sounds we were hearing were a rocket or not. Then our kids came into the bedroom and we covered them with blankets. We started reading messages and news. While calming the children, we also explained to them that a war had started and that we were going to get them to a safe place.

We understood that the battle for Kyiv would last long so we decided to move the kids as far away from rockets as we possibly could. Within hours, a huge number of people in cars began to flee the city and created many traffic jams. We waited and visited some friends and checked on others by phone. By 9:00 p.m., the traffic was much better so we left Kyiv for Lviv in the west. The trip took 26 hours instead of the usual 6 or 7.

After two days, I came back to Kyiv with a friend who is now in public affairs for Ukraine's army reserves, the Territorial Defense Forces. Svitlana and the children crossed the Poland border and were picked up by our friend in Germany, where they still live. I remained in Kyiv, keeping my computer reseller and consulting business running. I slept on a mat in the hallway for many nights because that was the safest place for me. I would also spend as much time as I could helping volunteers and networks of makers. For those of us not fighting the war, we helped in any way we could, connecting to other people and finding resources for them.

When we started producing Maker Faire in Ukraine in 2015, we hoped to organize a volunteer network of makers. We produced 15 Mini Maker Faires in five cities, a two-day camp for makerspace leaders, and had a plan to produce a maker-hacker camp for summer 2022. The goal is to change the culture, which Svitlana describes as characteristic of the Soviet era: "very low trust, low initiative, low ideas rate." She believes that Ukraine is "very hungry for ideas, startups, new forms of education, especially because schools and teachers are still old school."

DEFENSE EFFORTS

War is not something new for Ukraine. We have been in a state of war since 2014 and we were confident that a big war was coming. Our country

A Kyiv shop burning near a makerspace after a rocket hit

Volodymyr Babii, Yuri Vlasyuk

has been in a weakened state, and many of our institutions were influenced by Russia. What was necessary was a huge volunteer movement to help support the defense of our country by the armed forces.

When the war began, many of us decided to take things into our own hands, if just to overcome the fear and confusion caused by the invasion and lack of information. What emerged was a self-organizing effort, not dependent on our institutions. As makers, we can respond to requests from relatives in unoccupied regions, or from friends in the Armed Forces of Ukraine (AFU) or the Territorial Defense. We can use the tools we have or reach out to people in the network who have the competence or tools to help out.

Last fall, we conducted a survey of 196 Ukrainian makers and found their most urgent need was new training, and their strongest interest by far was 3D printing. There are now about 18 makerspaces across Ukraine, many which have struggled for funding during the war. A makerspace in Kyiv was hit by a rocket in November 2022; we lost equipment such as 3D printers but no one was hurt when the explosion hit in the early morning hours.

A large number of makers were drafted into the army where most of them keep improving processes, developing and iterating new tech, building trust in their team, and caring a lot about what they do. Civilian makers are helping to supply our defense with much-needed equipment as well as modifying or repairing devices.

Spikes welded at Kyiv Hacklab

Artist Sergiy Vukolov, with a small team, welded these camp stoves.

A finished camp stove

Artem Synytsyn, Sergiy Vukolov, Ihor Kochet, Oleksandr Panforov, Ada Yelagina, Jury, Keep Robotics

WELDING

At Kyiv Hacklab, residents welded a lot of anti-vehicle spikes — the metal comes from cutting up shopping carts — as well as Czech hedgehogs and other obstacles to stop heavy armored vehicles.

Makers also started to weld simple, cheap camp stoves and sent them to soldiers as well as civilians in unoccupied territories. Odesa artist Sergiy Vukolov raised some funds for supplies and then started welding camp stoves without a day off.

ELECTRONICS

Quest organizer, steampunk fan, and maker from Dnipro, Ihor Kochet, started to produce power supplies in ammo boxes using car batteries and sent them to the front line where they needed generators. A maker who I can't name started small-batch manufacturing of $100 low-power encrypted communication radio stations.

3D PRINTING

With supply chains broken, there was a need for ordinary car parts for repairs. Makers are often able to 3D print replacements.

3D Printing for Ukraine coordinated efforts outside Ukraine to print parts for tourniquets for wounded soldiers. Such tourniquets were not available early in the war. Another example is Tech Against Tanks in Warsaw, Poland, which has a database of 3D printing projects that includes eye shields, knee guards, and window barricades.

3D printers are also used to produce grenade-release mechanisms for drones and stabilizing tails for grenades. These tails are printed all over the country in huge numbers. Early in the war, the grenade tails made noise when used, which alerted the Russians, and so the community prototyped new versions that were quieter. They also had to correct for the grenades tumbling out of control.

This March, I organized RepRapUA, a festival to celebrate and mobilize our community of 3D printing entrepreneurs, suppliers, manufacturers, and enthusiasts. It was a huge success, with 901 visitors; about 30% were defense-related. For security, a secret location was necessary.

A car battery in an ammo box can be used as a universal power bank for devices.

3D printed car part (in blue)

Various 3D-printed grenade tails

Drone with 3D-printed shell to carry payload

DRONES

At the end of summer 2022, there were more than 200 different new drone projects that were a response to the invasion.

At the second Kyiv Maker Faire, we had established FPV drone shows and competitions. From the first days of the war, people from this drone community started to use their skills to fly standard consumer drones to understand battlefields, and then switched to professional units. Now they produce a lot of FPV-kamikaze drones.

Consumer drones were outfitted with 3D-printed grenades. Jury is an engineer who lives near the front lines and despite power outages, keeps doing this work and connects with the Armed Forces daily. He repairs drones and night vision goggles, and develops new mechanisms to release grenades.

Two ex-hackerspace engineers, Pavlo Shelyazhenko and Mykola Palamar, started Keep Robotics (en.keeprobotics.com) to build

Keep Robotics ground drone

Last-mile cargo bike by Kyiv maker Danylo Braverman

Overhead view of Tolocar mobile makerspace van with its contents

a multipurpose electric ground drone. They modify agricultural equipment to build a defense platform for evacuation, intelligence, and delivering ammo.

IMMERSIVE TRAINING

Before the war, Serhii Nezhinsky was engaged in a rare profession for our country: a digital curator. As co-founder of X-Platform and founder of the METΛCVLTVRE project, he worked to integrate new technologies into business and creative industries. "I believe that culture is the first line of defense of the country," said Serhii, "and that virtual art can be transformed into absolutely real security."

In 2022, Serhii volunteered for the army and now works in the Ministry of Defense where he is engaged in planning combat training and after-action review. Serhii is working on a dream project. "It is an immersive media (VR) training ground with a system of interconnected simulators of weapons, equipment, and aviation. That will make it possible to train on real locations, scanned to the smallest detail, and to plan operations in real time." The project has the potential to serve as a magnet for more creative talent like Serhii in the army.

HUMANITARIAN EFFORTS

Makers have also been involved in rebuilding efforts in de-occupied territories and addressing the needs for shelter, heating, light, and food for the people who live there. A lot of the work is helping more people learn new skills.

CARGO BIKES

Maker Danylo Braverman started to produce cargo bikes for last-mile food delivery to unoccupied regions where infrastructure was damaged. These bikes helped locals deliver food, stoves, and clothes in the Chernihiv and Sumy regions.

TOLOCAR: MOBILE MAKERSPACES

Toloca is an old Ukrainian word meaning "communal work." In early spring 2022, we developed the Tolocar project (tolocar.org/en) with funding from the German government's Agency for International Cooperation (GIZ) and deployed a fleet of four mobile makerspaces for humanitarian relief. These vehicles look like STE(A)M Trucks in the USA. They visit makerspaces and technical colleges, bringing equipment and tools as well as teaching workshops.

"Such workshops most often involve people (kids and adults) who have never done any making before," explained Olga Ivanchenko, who works with a small team in Kyiv to operate the mobile makerspace and teaches 3D printing. "In this way, we try to show them new opportunities for technical creativity and interest them in science. We think that crafting has a great positive effect on the psyche of people and can distract from the horrors of war, give impetus to new activities, and strengthen the economy of our country."

Because of frequent blackouts, lighting is a big concern. "We developed a small, powerful flashlight for emergency lighting," said Olga. A workshop at Hacklab Kyiv walked people through the flashlight project. "Together with volunteers, we assembled emergency lights for three different animal shelters and then we started going to women's shelters and organizations supported by the International Medical Corps and doing emergency lighting for them," added Olga. The flashlight is an open source project at: github.com/tolocar-project/solutions/tree/main/Emergency%20light. The people who took the classes said that it helped them feel more capable, creative, and supported by the community.

Olga along with Volodymyr Babii have worked with various organizations. They developed a heating efficiency upgrade and made an emergency power system for Hacklab Kyiv. They worked with the Bobry (which means "beavers" in Ukrainian) workshop in Sumy to make a spot welding machine together so that they could independently weld batteries and make power banks.

Tolocar can help people feel like someone really cares because they invest time, share skills and knowledge, and stay in touch afterwards. I also hope that it will help people recognize the value of having such a space in their community to get additional support and training.

NEW SKILLS TO SURVIVE

Learning new skills is necessary to survive or to stay in business. Julia Yalanzhi runs Yalanzhi Objects, a small manufacturing business in Kyiv that produces eco-friendly lamps. She says her project is a "combination of art, handicraft, and utility using ceramics and recycled paper." Because all the men who worked in her business were mobilized by the army, she took it upon herself to learn how to weld and to solder. "The war tore our family apart," said Julia. "My husband went to the AFU, and I went into manufacturing. New times forced me to learn new skills."

Olga Ivanchenko teaches a 3D printing workshop in Chernivtsi.

Success: working flashlights

Julia Yalanzhi learned to weld at Hacklab because her workers left to fight the war.

Volodymyr Vetrogradskiy at Hacklab Kyiv

Yuri Vlasyuk, Artem Synytsyn

As for my family, Svitlana remains in Germany with our two children and I remain in Kyiv. The war is lasting much longer than we expected and there is no clear path for it ending. We will continue to have a huge need for small batches of civilian and defense products, for makers and small industries to come to the rescue to help solve problems quickly with the help of innovative technologies. We will continue to need makers.

UKRAINIAN MAKER ASSOCIATION

Since 2015, Svitlana and I have worked under Maker Hub NGO. Now, it is time to gather all the beneficiaries and develop a platform to support makerspaces and makers all around the country. Late last year, we established the Ukrainian Maker Association (makerua.org) to raise funds and build structures to help makerspaces and makers — those that are already established and other spaces that are starting up.

WHAT AMERICA CAN LEARN FROM THE DEFENDERS OF UKRAINE

BY BRAD HALSEY

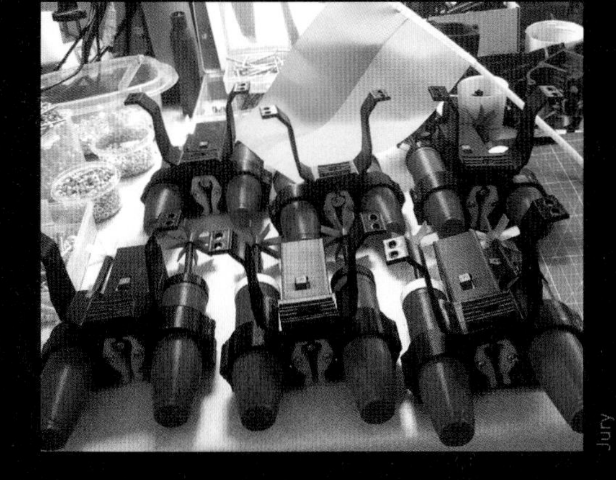

Jury

In my recent trip to Ukraine I was astonished to see that the primary driver of new innovations in the conflict was entrepreneurs and small business owners. As an entrepreneur, I know pivoting and creative problem solving is the only way to survive. So of course that mindset is helping the Ukrainians combat the Russians.

I witnessed former start-up CEOs-turned-drone-creators 3D print mechanisms designed to drop mortars onto Russian tanks using drones they learned how to build using YouTube ... and I wonder if our military would allow such civilian participation if ever a war were to come to us?

All Ukrainians are fighting against the Russians, not just the military. Everyone feels compelled to help in some way. Churches have become distribution points for everything from blankets to ammunition. Taxi drivers are shuttling fighters to the front lines. And Ukrainians good at learning and problem-solving quickly — entrepreneurs for example — are developing really clever technologies using just what they can find on the internet.

Everyone has seen the effectiveness of drones in Ukraine. The cleverness is definitely driven by necessity. Modifying and testing drones can happen in a matter of hours, not weeks or months. The local countryside becomes the test range for new concepts and novel bomb-delivery without onerous red tape and bureaucratic crap, something our military has not figured out.

We should learn from Ukraine. Flying drones is increasingly difficult in the United States and nearly impossible on military ranges — the same ranges where rockets, mortars, and machine guns are fully permitted. As with every overreaction to new tech, the problem is rooted in lack of understanding. How could a 700-gram piece of slow-moving plastic with no explosive capability be more dangerous than the bullet of a sniper rifle, yet the regulatory process for the flying plastic is stifling?

Yet, in any future conflict, drones will be extremely useful as they certainly are in Ukraine. And our military will be racing to show every warfighter how to operate and modify drones. So why don't we embrace that tech and encourage use before such a situation exists? Let's watch and learn from the Ukrainian people and mix a little common sense into our bureaucracy, especially with new technologies.

Unfortunately, I fear we will continue to overregulate and constrain new tech until our military gets faster at understanding that the dangers are not proportional to the benefits. AI tools like ChatGPT will likely be next in this fear-driven paralysis of adoption.

Ukrainians are learning how important problem-solving with tech can be while fighting for their lives — let's ensure we learn how to effectively use tech *before* it becomes a necessity to do so.

BRAD HALSEY is CEO and founder of Building Momentum in Arlington, Virginia. He is passionate about solving humanity's hardest and most urgent problems. Sometimes this means dropping into war zones and disaster areas to rapidly develop solutions. Brad visited Ukraine last December. He wrote "Fighting Disasters" in *Make:* Volume 61.

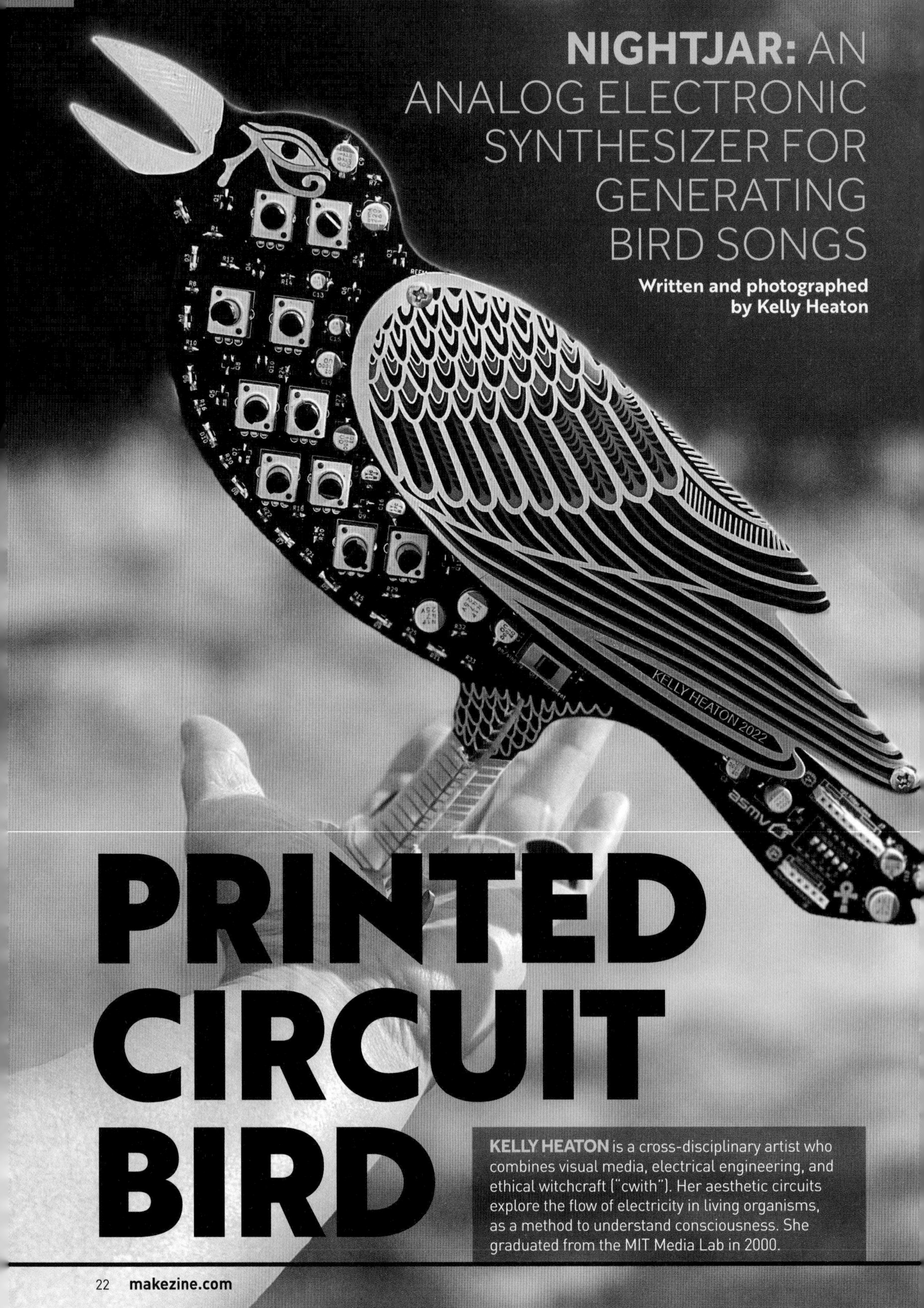

NIGHTJAR: AN ANALOG ELECTRONIC SYNTHESIZER FOR GENERATING BIRD SONGS

Written and photographed by Kelly Heaton

PRINTED CIRCUIT BIRD

KELLY HEATON is a cross-disciplinary artist who combines visual media, electrical engineering, and ethical witchcraft ("cwith"). Her aesthetic circuits explore the flow of electricity in living organisms, as a method to understand consciousness. She graduated from the MIT Media Lab in 2000.

Nightjar is an electronic sculpture that generates bird-like sounds. The circuit is completely analog, which means there's no digital code or recorded audio, only hardware vibrating in the audible range.

Nightjar represents many years of breadboard experimentation during which I developed my practice of *electronic naturalism*. As Nikola Tesla famously said, "If you want to find the secrets of the universe, think in terms of energy, frequency, and vibration." I followed his advice to study electrical patterns in living and life-like entities, and Nightjar is my greatest engineering accomplishment to date.

Nightjar's design is remarkably simple. The circuit sings like a bird with only six astable multivibrator oscillators, several passive filters, and an 8Ω speaker — all powered by a 9V battery. Colored LEDs reveal how electricity vibrates in the circuit. Ten potentiometers adjust resistance to alter the dynamically generated audio signal and produce different artificial bird songs. It's an instrument for "playing nature," perhaps even the start of a new musical genre.

Kelly Heaton with Nightjar during her artist residency in Integrated Design & Media (IDM) at NYU Tandon School of Engineering, Spring 2023

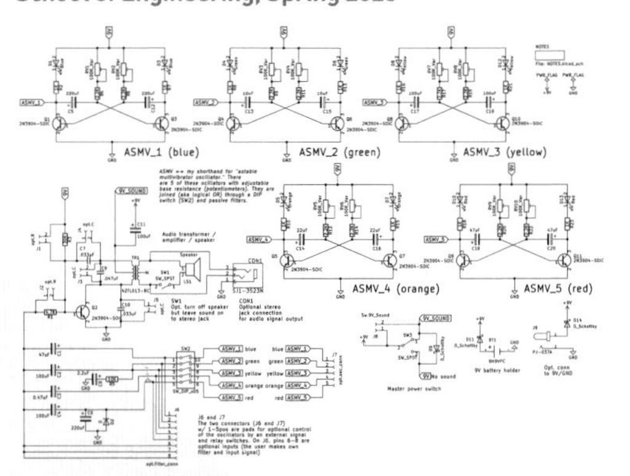

Prototype schematic for Nightjar, 2023

FEATURES FOR MUSICIANS AND MAKERS

Nightjar has an audio line-level output so that musicians can literally go wild with it. Plug it into your favorite synthesizer to make naturalistic soundscapes, angry squawking, or bird songs from outer space. Makers can use the output signal to control video or something else entirely. I'm working on manufacturing and distribution to make it available for purchase later this year.

Nightjar's five main oscillator connections can be mechanically switched on or off with its DIP switch, or remotely controlled with your own keyboard or relays. There are three inputs to Nightjar's voice mechanism (a modified Hartley oscillator) where you can introduce any signal up to 12 volts. The voice quality is set by capacitor values affecting pitch, so try modifying these components if you want create a unique sound. Or if you're in a quiet mood, set Nightjar to "lights only mode" and enjoy a silent electronic sculpture.

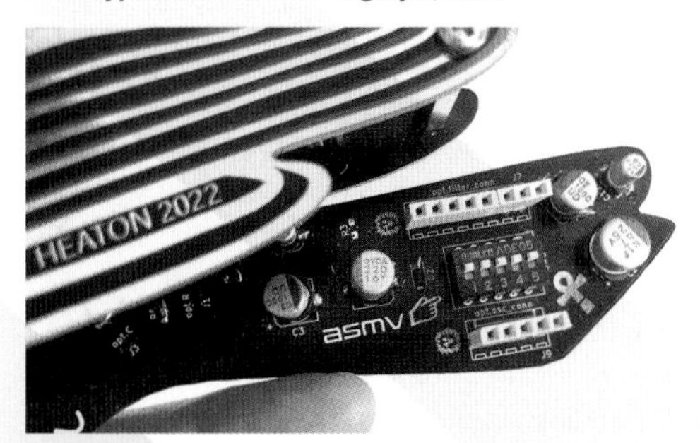

Detail of Nightjar showing the oscillator DIP switch, as well as pin headers for optional remote control, and three slots for additional audio inputs

A CURIOUS AND DETERMINED BIRD

Once upon a time, songbird circuits were far beyond my technical ability. I began my journey

Luke DuBois connecting Nightjar to a Serge synthesizer for the first time. IDM, NYU Tandon, 2023, vimeo.com/804078922

***Dot-winged Ant Wren*, 2018. Freeform electronic circuits and watercolor on paper, 15"×10"×1". Created while studying birds in Akumal, Mexico.**

***Breadbird*, 2019. Analog electronics and mixed media interactive sculpture for birdsong generation. Used for the *Deep Fake Birdsong* experiment (2020).**

with electronic art in the late 90s by hacking toys such as Furby and Tickle Me Elmo, and later developed my own circuits to depict the flow of energy in nature. I focused on electronic hardware (as opposed to software) because I wanted to understand the fundamentals of electricity as a creative medium.

My earliest sound generators were chirping insects. I was determined to make a circuit sing like a bird because, while insects are cool, birds have a next level of vocal intelligence. There was no instruction manual for animal sound circuits, so I scavenged hobbyist schematics and tweaked them; the Mini-Notebooks of Forrest M. Mims (see pages 82 and 100) were a favorite resource.

My earliest "circuit birds" were freeform electronics built into watercolor paintings on paper. Later, I sculpted the hardware into an electrical anatomy that's visible inside of transparent birds. In 2019, I created my first "printed circuit bird" as a limited-edition kit called *Pretty Bird (ver.CC)* and donated 120 of them to raise money for the arts. Their popularity led to several commissions, including my recent *Circuit Garden* (2022), which showcases my electronic menagerie in a playful landscape.

DEEP FAKE BIRDSONG

Nightjar's earliest direct ancestor is *Breadbird* (2019), which has all the same design elements, but in handmade form. In early 2020, I asked Johann Diedrick to use his artificially intelligent software, Flights of Fancy, to analyze the sound generated by my birdsong circuit. What would Diedrick's algorithm think about my artificially intelligent hardware? Would the software detect a real bird? What species? The experiment was earnest, but also a sort of electronic Dadaism.

Diedrick's software matched 47 of the 122 analyzed spectrograms with *Antrostomus sericocaudatus* (aka the Silky-tailed Nightjar) with a 93% average confidence rating ... and so the name Nightjar stuck!

ELECTRONICS MODELING LIFE

For more about the parallels between analog electronics and life, check out William Grey Walter's *Machina speculatrix*, David Dunn's *Mimus polyglottos*, and Mark W. Tilden's BEAM robotics. If you have other examples, please share them!

MORE GOODIES

- **Kelly Heaton Studio:** kellyheatonstudio.com
- **Instagram and Mastodon:** @kelly_heaton
- **Nightjar available for purchase from Adafruit:** adafruit.com/product/5654
- **Hacking Nightjar:** vimeo.com/779394321
- **Nightjar and Serge:** vimeo.com/804078922
- **Electronic naturalism:** kellyheatonstudio.com/electronic-naturalism
- ***Deep Fake Birdsong*, 2020:** kellyheatonstudio.com/deep-fake-birdsong
- **Hacking Nature's Musicians, 2018:** hackaday.io/project/161443-hacking-natures-musicians

YES IT'S LEGAL TO BUILD AND SHOOT YOUR OWN — WITH HELP FROM THE **DIY FIREWORKS COMMUNITY**

Written by Victor Chaney

VICTOR CHANEY is a dentist and lifelong maker who has been making fireworks for 4 years. He has written for *Nuts & Volts Magazine* and Instructables.

MAKING FIREWORKS

Make fireworks? That sure sounds like fun! But isn't that illegal in a lot of places? No problem! Go to a fireworks convention, where you can take classes. I got started at the Western Winter Blast, held annually by the Western Pyrotechnic Association. There, with about 800 other people, I learned how to make fireworks and shoot them in a safe, desert environment. I also saw fireworks and shows that dwarfed anything I'd ever seen before — amazing pyrotechnic works of art. You can spend hours making something that's gone in 10 or 20 seconds, but what a glorious 20 seconds it is!

WHICH FIREWORKS CAN YOU MAKE?

The foundation of many firework creations is the *ball shell*. A spherical shell of papier-mâché holds the *stars*, which are the bright points of light that make the colored trails. The shell is fired from a mortar, or launch tube, where a *lift charge* of black powder sends it on its way in a predictable, pre-determined path of travel. Stars can be placed in multiple layers, or *petals*.

Cylindrical shells, aka *Italian shells*, can be greater than 12" in diameter and 3–4 feet or longer. A *shell of shells* has several small shells inside a big one.

Rockets are self-propelled and often explode at the top of flight into stars or *salutes*, which are loud bangs. They go up with a whoosh, and while often smaller in payload than shells, the takeoff and burst make a dramatic display. They are less predictable in their path; some shell enthusiasts refer to rockets as "the dark side" of pyrotechnics.

There are so many other kinds of fireworks like *mines*, *comets*, *gerbs*, *Roman candles*, *fountains*, and the spectacular *girandola*, rimmed with rockets, that goes up in a spinning wheel of glittering, noisy spiral trails.

BRINGING THE BOOM

At the heart of most fireworks is *black powder*, made of three ingredients. Potassium nitrate (KNO_3), is the oxidizer that provides oxygen for the reaction. Charcoal is the fuel, and sulfur is a fuel that serves to speed up the reaction.

Technically, black powder does not explode, but burns very fast. The shell is wrapped tightly to build up a lot of pressure so that it bursts with great vigor. A dash of *slow flash booster* (KNO_3, sulfur, and aluminum) gives it some extra pop!

Fine black powder coated onto rice hulls is called *burst powder* and is used to fill up the inside of shells. Granular black powder, grade 2FA, is used for the lift charge to launch a shell. It is also at the heart of stars, rocket fuels, and fuses.

Stars are typically composed of an oxidizer, a fuel, a binder, and something to make it shine. Copper compounds give a blue color, strontium gives red, and barium gives green. Metal powders make sparkles— typically aluminum, titanium, magnesium, and iron.

HOW A SHELL IS MADE

For a double-petal ball shell, you start with two *hemis*, or halves, of papier-mâché or plastic. One has a *passfire tube* glued into it, which will allow the fire to get into the finished shell. The hemis are then lined with a layer of stars, a layer of tissue paper, burst powder, more tissue, and then the second layer of smaller stars to form the inner petal. I made a 3D-printed jig to help place these. The center of the shell has a final sphere of burst powder inside.

The two hemis are put together and wrapped with layers of gummed paper tape to give a dense outer layer that allows plenty of pressure to build up inside before the shell bursts!

» **FREE DOWNLOAD:** To learn more about DIY fireworks, conventions, tools, and safety, get the complete "Making Fireworks," by Victor Chaney and Ellen Webb, at makezine.com/go/fireworks.

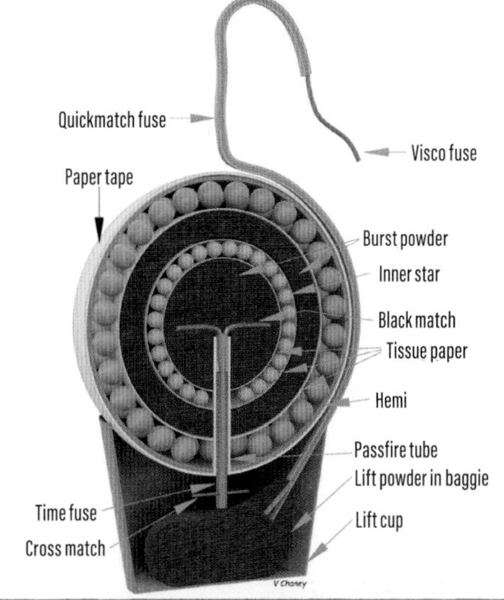

DOUBLE-PETAL BALL SHELL DIAGRAM

Stars are placed in a hemi for a large shell.

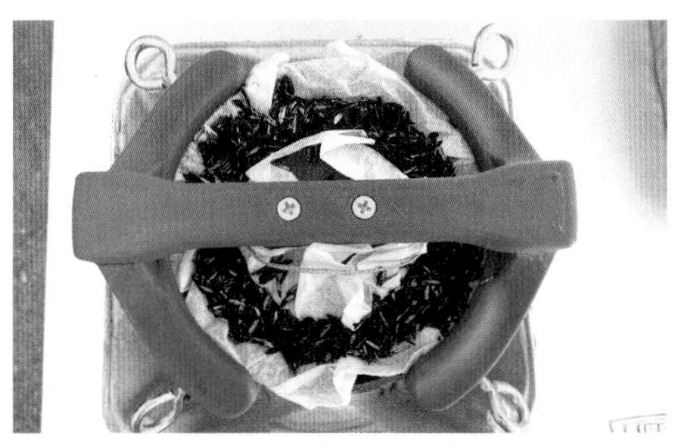

A 3D-printed jig allows compaction of burst powder over the outer petal of stars, saving space for the inner petal.

Wanda Garrett compresses fuel into a rocket with a large press.

Victor Chaney, Mike Garret

TO BRING A *THING* TO LIFE

Written and photographed by Jesse Velez

Building the "impossible" robot hand

JESSE VELEZ is a designer and fabricator, and co-founder of Raptor House FX, creating specialty items for film, television, live entertainment, and a variety of creative industries. He is crazy about movies, science fiction, and making in all its forms.

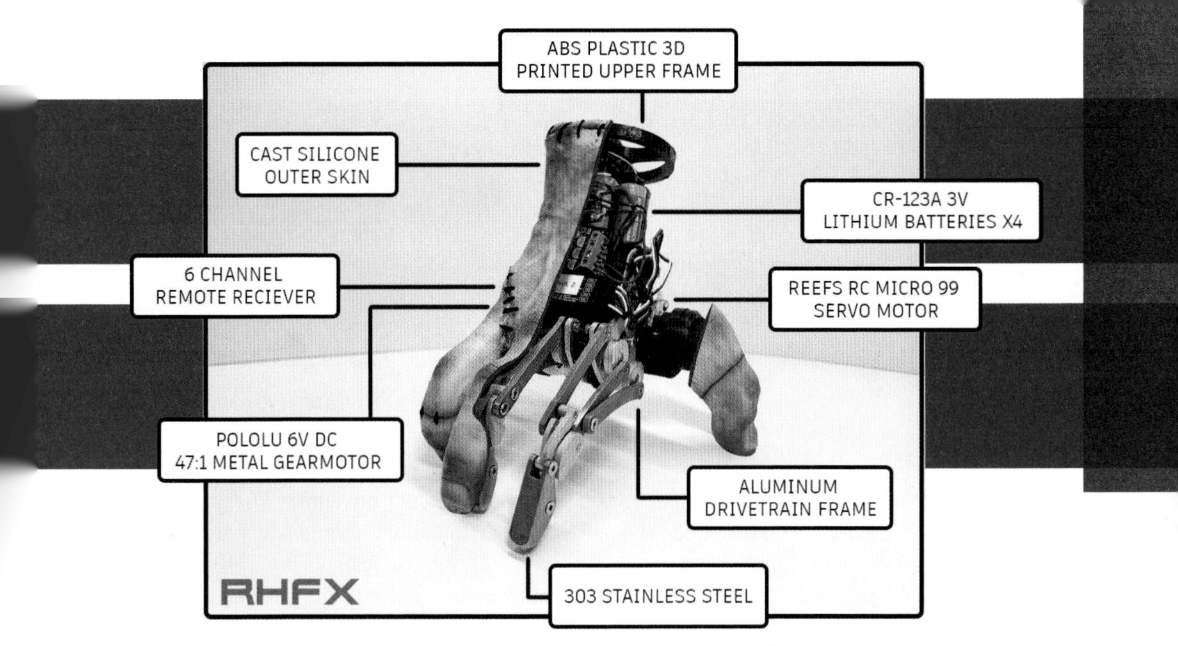

In April of 2022, our special effects studio Raptor House FX (RHFX) was approached by Netflix to explore the feasibility of creating Thing — the classic Addams Family disembodied helping hand — as a self-contained, walking, animatronic puppet.

The project was conceived by the viral marketing agency Whoisthebaldguy to help promote the now critically acclaimed series *Wednesday*, directed by Tim Burton. Videos intended for social media would follow Thing as he interacted with the unsuspecting public of New York City and Los Angeles; maybe he would even appear at red carpet events and premieres.

The aim was to produce a mechanical magic trick not yet seen in any of the previous Addams Family permutations. For more than 50 years, Thing has been played by a human actor or hand model, originally hidden inside props or furniture, and subsequently "painted out" with green or blue screen compositing technology.

The latest version of Thing that we see in Wednesday was created in this same method. Actor and illusionist Victor Dorobantu brought new life to the character, now redesigned to appear as a Frankenstein-esque, stitch-covered, living severed hand with attitude. On set, Victor was dressed in a chroma blue leotard, with makeup effects applied to his exposed right hand, and a prosthetic wrist stump serving as a clever visual break to further sell the illusion.

To bring a Thing to life practically, without a human actor, would require a wholly unique and surprisingly complex radio-controlled robot (Figure A). I met with Netflix in April and what followed was an intense 5-month marathon of R&D, from proof of concept, to fine tuning and testing before the shoot, to final aesthetics.

The first hurdle for our team at RHFX was simply to define what we were attempting to create. Thanks to the Addams Family, the sight of a crawling human hand is a common fictional image, but in reality human hands do not walk, or even support their own weight! Nothing about the evolutionary design of hands lends itself to ambulatory movement. The physical form of a human hand, with five legs (fingers), topped with an unbalanced wrist mass, required a completely original approach to animatronics design.

While the final product needed to resemble a human hand, the team was in fact not designing a hand at all; rather, they were designing a pentapod (5-legged) robot with uneven leg placement and a flesh-like silicone skin.

A

We explored various design philosophies including motorized joints, wheels hidden in the fingertips, and flexible camshafts; but in the end we landed on a single DC drive motor, accompanied by three servomotors for secondary animation. The central issue was designing a fixed mechanical linkage for the pointer, middle, and ring fingers that would be capable of pulling the hand forward across a variety of surfaces.

To solve this puzzle, we hired Canadian maker Ben Eadie, engineering technologist and movie special effects designer [see "Next Level Radio Control" on page 31]. Pulling inspiration from the well-known *Strandbeest* by Dutch kinetic artist Theo Jansen, Ben designed a linkage system in OnShape for each finger comprised of three rigid, curved "beams" that pivot on one fixed upper point. The second and third beams are attached to rotating cams, driven in fixed ratios to one another by a tightly packed gear stack, which is powered by a toothed belt and a DC motor. Once the finger timing was set, each component could be locked in place relative to each other, and a dependable pulling or pushing force could be generated to move the robot forward or back. Additionally, I designed articulating fingertips, which when joined to Ben's upper leg linkage, created a lifelike "flick" at the end of each step.

The first tethered walking proof of concept was finished in July, which green-lit the final phase of design. Updated versions of each component were modeled and sent out to various CNC milling shops for speedy manufacture.

The final robot was designed with a bottom-heavy "mass gradient": the fingertips and leg linkages were machined in 303 stainless steel, while connecting "tendons" and the gear case frame were cut from lighter 6160 aluminum. Atop all that, the battery frame and motor housing are 3D printed in ABS plastic. The result is an extremely stable standing position that can rock back and forth on pinky and thumb servos to achieve lifelike character animation.

All in, the final mech contains close to 50 components, each designed from scratch and either 3D printed in-house at RHFX, or machined by our fabrication partners. The finishing touch: real stitches to close up the silicone skin. Two identical puppets were built so that any on-set repairs would not slow down production.

The final result was a lifelike walking hand which delighted and frightened civilians when it premiered on the streets of New York City (Figures B and C). RHFX definitively proved that the character could be brought to life with animatronics. Children screamed, horror fans laughed, and everyone asked with amazement: "How did they do that"? Netflix executives and Tim Burton himself were thoroughly pleased with the result, and the videos produced have received upwards of 100 million views worldwide (see for yourself at youtu.be/B479Wc72Bsc).

Wednesday has become one of Netflix's most watched series, with a second season green-lit for production. If the stars and budgets align, we may yet see a practical animatronic Thing crawl on screen sometime in the near future.

- **Project lead** — Jesse Velez
- **Walking mech** — Ben Eadie
- **Workshop assistant** — Miles Berwick
- **Moldmaking and skinning** — Cali Jones
- **Silicone paint and finish** — Mariah Kierns
- **Machinist** — Chris Mora

NEXT LEVEL RADIO CONTROL

Bring models to life with channel mixing, telemetry, and on-the-fly programming — using OpenTX for your R/C transmitter **Written and photographed by Ben Eadie**

Radio control is something I use in the creation of special effects and props in movies and TV. It is now one of the things I am known for. But five years ago, I had very little idea how to make these transmitters and receivers work!

A little curiosity sprinkled with some creativity can really up the game on what you can do with radio control, and the wow factor you can get. Progressing from a beginner to an advanced user is nowhere near as hard as you might think. I encourage you to dive in and try some of this stuff.

MATERIALS

- **Radio control transmitter, OpenTX / EdgeTX compatible** For a list of compatible transmitters see www.open-tx.org/radios.html or edgetx.gitbook.io/edgetx-user-manual.
- **R/C receiver** to match your transmitter
- **Motor controllers**
- **DC motors, servomotors, lights, etc.**

TOOLS

- **Computer with OpenFX Companion and/or EdgeFX Companion software** free and open source from www.open-tx.org or edgetx.org
- **USB cable** to update your radio's firmware

LINUX TRANSMITTERS

The first thing I would recommend for people to do is to look at alternative radio systems past the standard R/C transmitters you see at a hobby shop, such as Futaba and Spektrum. Now, these are great transmitters; if you want plug-and-play, basic functions, these will be the ones you should have.

However, if you want to get into some advanced channel mixing, telemetry, and a system that is easily programmed via your computer, you need to be looking into radios such as **FrSky**, **Jumper**, and **RadioMaster**. These don't have to cost more than your standard hobby store radios either. You can get a decent 8-channel radio for $120 like the Jumper T-Pro. Or go all out and buy something like the FrSky Horus X12S worth $600 or more for all the bells and whistles.

OPEN SOURCE FIRMWARE

All of these transmitters run on open source Linux-based operating systems. One is **OpenTX** (www.open-tx.org) and the other is **EdgeTX** (edgetx.org), a fork of OpenTX that's mainly used for developing new features. Both are very similar and capable.

Why would you want one of these systems? Let's start with an example of how to implement it to get the movements you want from a Mecanum robot platform. Mecanum is a type of wheel system used for creating unique and fluid movements that standard wheels cannot. It's made up of several small rollers that are arranged at an angle on the wheel (Figure A).

When, for example, you spin the left-side wheels in a counter-rotating direction to the right side (Figure B), the robot can move sideways! A combination of sideways movement and forward will move it diagonally, or you can do your standard skid steer movements like a tracked vehicle (Figures C and D). Basically, a Mecanum-wheel drive can move in many different directions at once! This makes for a robot that can move in unique ways. To mix the channels on a standard transmitter to achieve this movement is possible — but honestly it's a huge pain. I like the path of least resistance. And that path took me to OpenTX and EdgeTX software running on the radios mentioned above.

OpenTX offers advanced features and capabilities to help you get the most out of your radio-controlled model. Effectively, it's a Linux system in a transmitter. You can program it to do most of what you'd need from a robot. This can make it so you no longer need an Arduino onboard your robot at all. You can offload all that work to the transmitter and do all the programming there.

These radio systems come with a computer

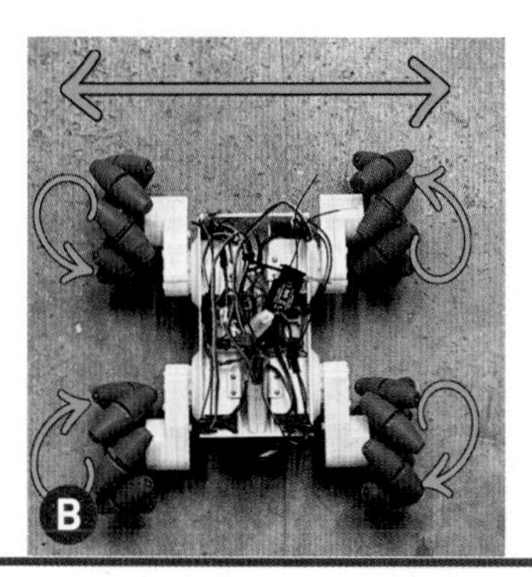

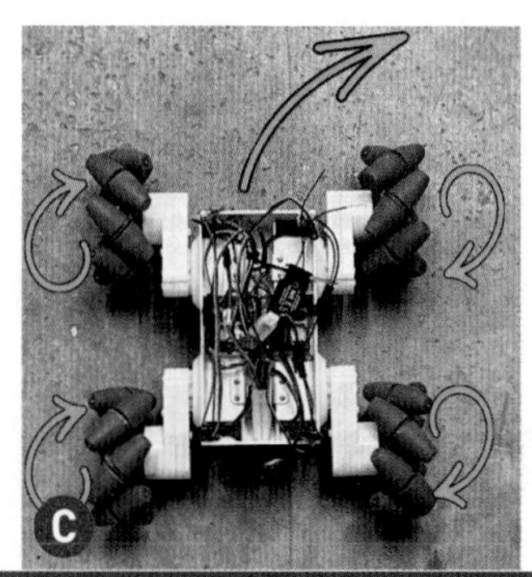

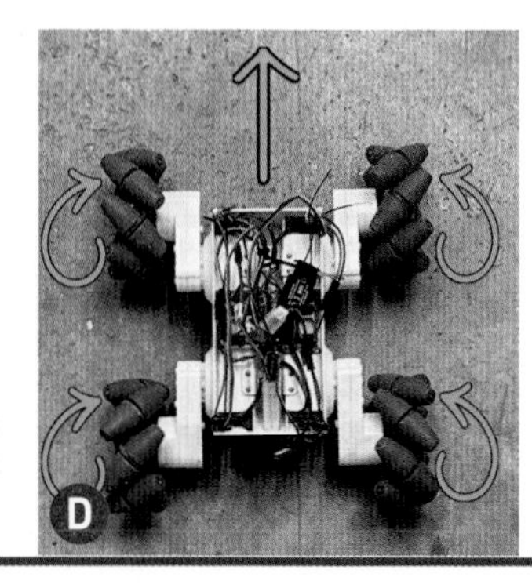

"companion" software that lets you plug your transmitter into the computer and program it (Figure E). This alone is a big step above standard radios. Programming a radio via a computer is amazing! Another great feature — you can run them as simulators, to test out your movements, before you ever send commands to the robot.

CHANNEL MIXING

Let's take a closer look at my preferred system, OpenTX, and how to use it for ***mixing channels*** — the secret sauce that allows you to control multiple movements by mixing inputs for your robot, model, or prop.

With our Mecanum robot platform, we attach the motor controllers to the receiver, each one getting a channel. I used channels 3, 4, 5, and 6 (Figure F). From there, we move completely to the transmitter for all the rest of this.

What follows below is the top-level look at how it's done. I'm not going into much detail, but the OpenTX and EdgeTX communities are very helpful, and you can figure this out via the guides on the websites or on the forums.

"Wilson Has a Mind of His Own"

Ben Eadie was working at Fuse33, a makerspace in Calgary, Alberta, where he met some movie people and got invited to work on special effects for *Star Trek Beyond*. That led to FX work on movies like *Ghostbusters: Afterlife*, which led to a friendship with a certain Mythbuster.

"I got a phone call at the beginning of the year from Adam Savage," Ben writes. "He wanted to know if I could build a sphere robot. Of course I said yes, even though I really had no clue. Because no one wants to disappoint a friend. Then he told me the robot was for Tom Hanks ... *the Tom Hanks!* Before I could eat humble pie, I was on the phone with Tom telling him that I could do what he asked. Follow along on my journey to find out if it all worked."

» **Tom Hanks Asked Me to Make a Robot Wilson for His First Pitch:** youtu.be/l5ps2VzAfA0

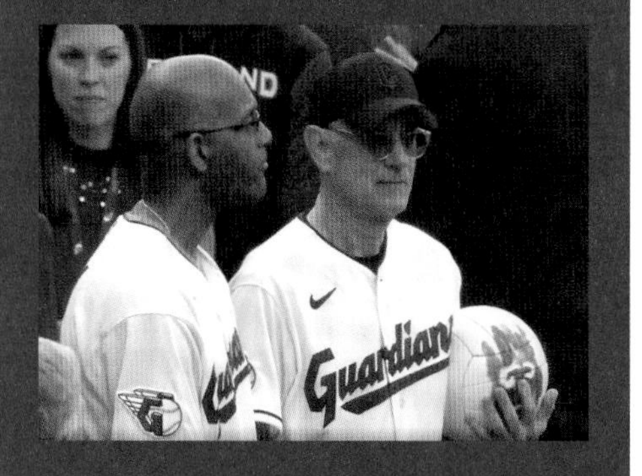

youtube.com/@DreadMakerRoberts

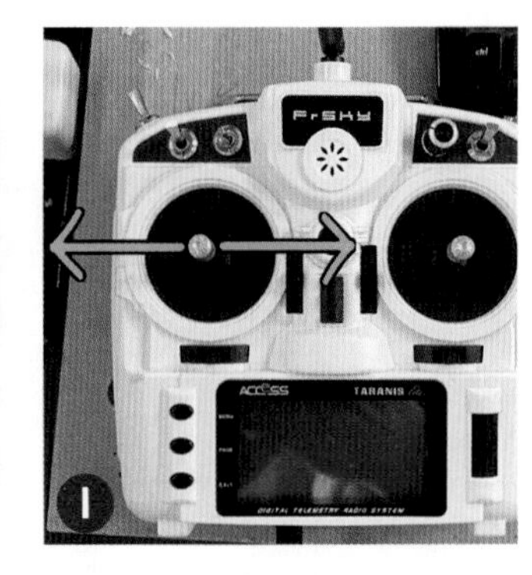

To start, decide which input channels (sticks on your controller) you want to use to handle the robot's motions. For forward and back motion, I chose the right stick's up and down movement (Figure G). Next, choose the turning motion; I chose the right stick again, but the left and right movements of it (Figure H). Lastly, the cool side-to-side sliding, aka "translate" motion; I used the left stick and its left-right movement (Figure I).

In the OpenTx Companion software, open the model you're working on (Figure J). Under the Inputs tab you can define your input channels (Figure K), in this case **FWD** for the forward-back movement from the right stick, **LR** for the left and

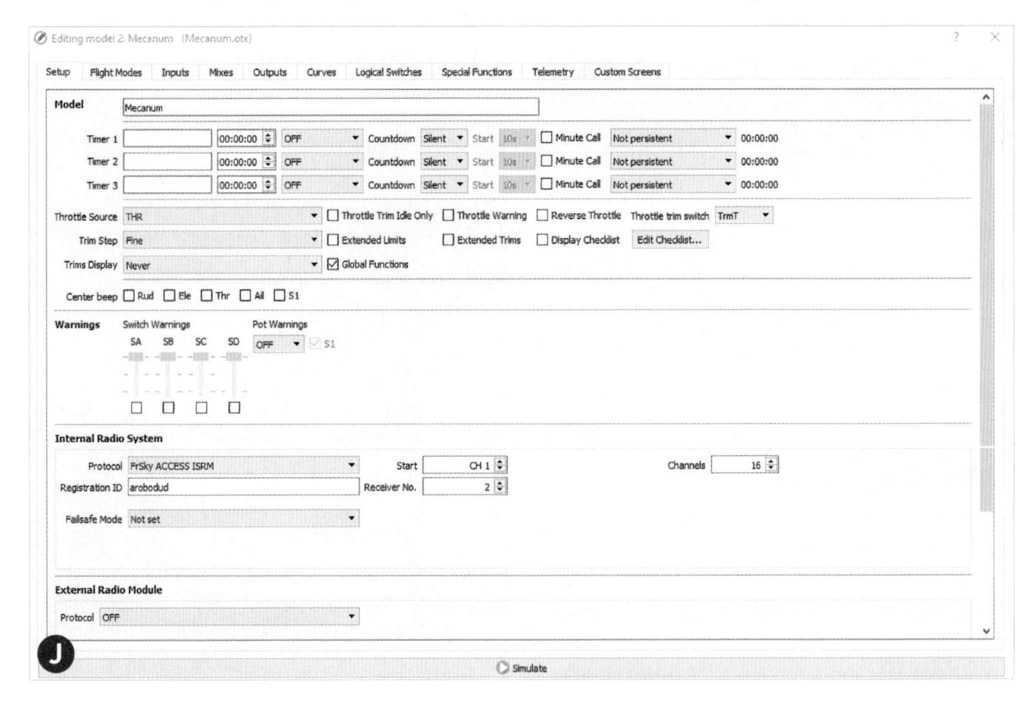

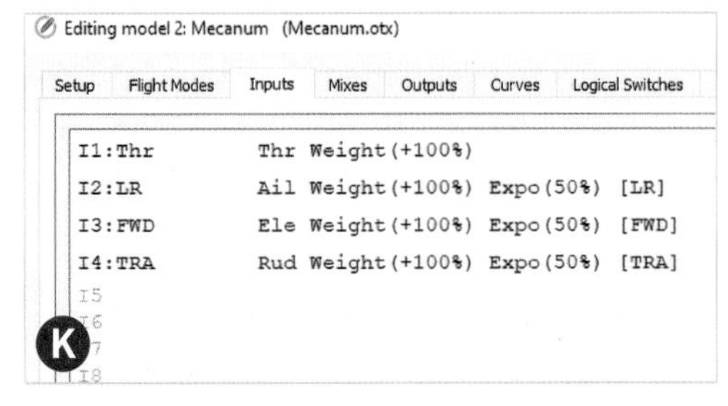

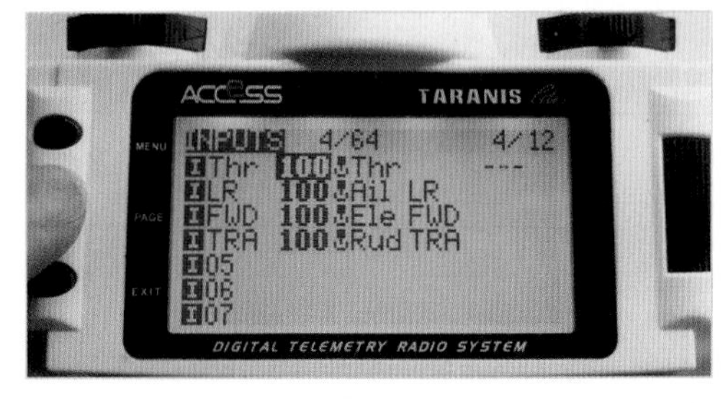

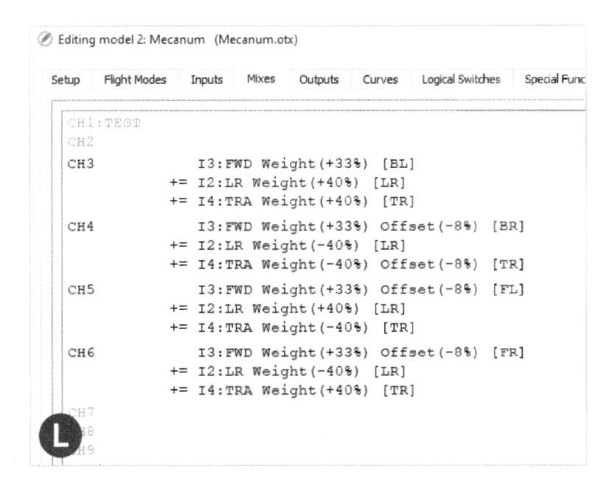

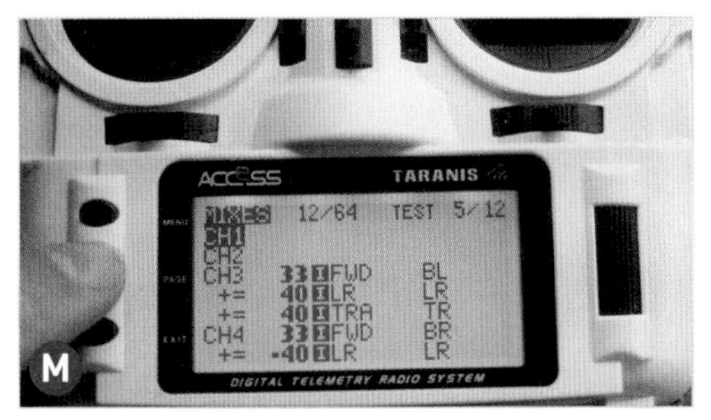

right turning movement from the right stick, and **TRA** for the left-right translation movement from the left stick.

This is where the Mecanum hits the road. With these inputs set, you can now mix them to allow you to combine inputs into some really cool movements on your robot. Under the Mixes tab (Figure L), you'll see the output channels you have connected your motors to on the receiver. This is where you map the inputs from the controller (Figure M) to the output channels the motors are on. You can mix each of these inputs from the controller here. Each channel can have a percentage of the input, or be on a curve of influence using a formula, or even be activated and deactivated via a switch! Here I have mapped each of the sticks, mixing a certain part of their signals to the corresponding output channel. Note how in some cases the input is subtracted instead of added to the output. These things can be a logic puzzle, and you need to really think of how one stick needs to interact with the other.

Following each input you will see a percentage, positive or negative, which determines the influence they each have on the output channel. For example, for the channel 3 output **CH3**, the front left wheel, you will see:

Who You Gonna Call?

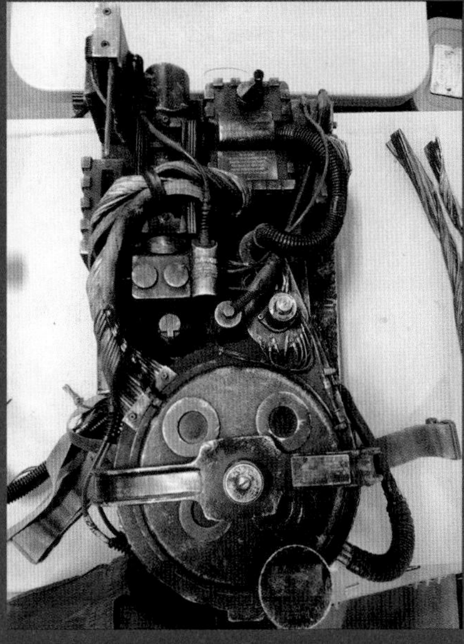

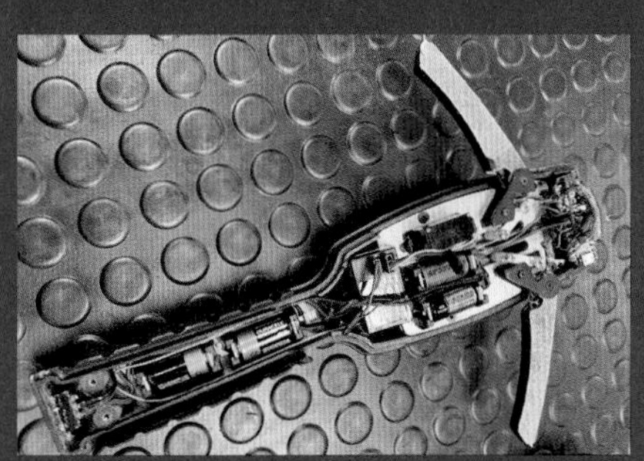

Next-gen ghost hunting gadgets from *Ghostbusters: Afterlife* (2021) — including Proton Packs, Psychokinetic Energy (PKE) Meters, and the new RTV (Remote Trap Vehicle) — are all radio controlled and built by an amazing team of SPFX and prop experts including Ben Eadie.

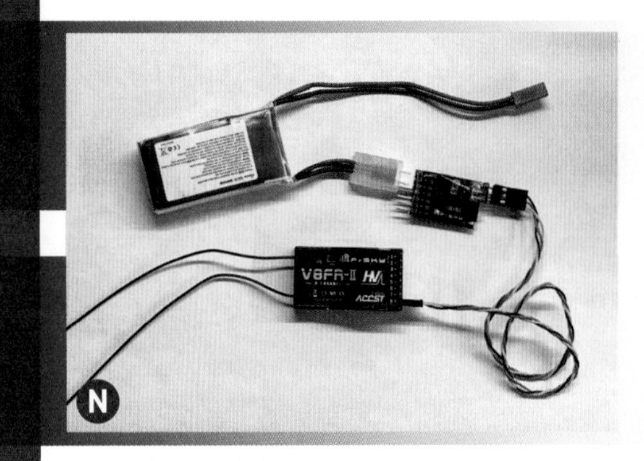

N

```
   I3:FWD Weight (+33%)
+= I2:LR Weight (+40%)
+= I4:TRA Weight (+40%)
```

Let's go through what this is telling you:

- `I3`, `I4`, and `I2` are all the inputs. `I` = input, and the number is the channel or stick
- `FWD` = forward-backward movement
- `LR` = left and right turning
- `TRA` = left-right translation
- `Weight` = the percentage of influence the input channel has on the output to the motor, controlling its speed and direction.

NOTE: In some cases, to get the right direction of spin, you need to *subtract* the influence and not add it; sometimes Mecanum wheels need a counter-rotation to do things like translate. In the end, it's either an exercise in logic, or you can figure it out by experimenting until you get the movements you want by messing with the percentages. The key to this is to *write it down* on each change you make so you know what worked and what did not.

REPROGRAM ON THE FLY

You may ask, "Why not get an Arduino to do this?" Well, for one, I find OpenTX much easier. I am not a coder, and the GUI of the companion software helps me understand exactly what is causing which movement.

Another big advantage of programming the transmitter instead of programming an Arduino or microcontroller is that having all the "brains" on the radio controller allows you to instantly customize and fine-tune your model's functions in real-time, without constantly re-flashing a chip onboard the robot. This level of control and flexibility is essential for unlocking the full potential of your robot — and for leveling up your newfound R/C guru skills. With the ability to reprogram on the fly, you'll have the power to make quick and effective changes as needed.

TIP: Try ChatGPT! There's a learning curve on these transmitter platforms, but I'm now using ChatGPT to write the programs and it works great! Try it out and tweak it from there.

TELEMETRY AND FEEDBACK

But wait, there's more! With these operating systems, you can get what's known as ***telemetry***. This is the ability for sensors like voltage, temperature, IMUs, barometers, and the like to send info from the robot back to the controller to inform you of what is going on or influence how the model behaves. You can even log data during a run to find out what is happening (and why) to the vehicle or the environment it is in.

You can have altitude sensors, flight speed, ground speed, and GPS on an aircraft. Temperature, humidity, cameras, and battery sensors on ground vehicles. A lot of these are also plug-and-play, meaning you plug them into the receiver, and they instantly feed this information back to your controller.

Here's an example. For this Mecanum robot I'm using LiPo batteries, and if you take them down past a certain voltage, they are destroyed forever or have been damaged to the point of uselessness. So, I have attached a battery sensor that will monitor the battery's voltage and also the charge level. Figure N shows the sensor plugged into the battery and into channel 8 of a receiver.

- This battery info is fed back to the remote (transmitter), which will display how much **power** I have left.
- With the companion software, I can then take this info and figure out how much **runtime** I

Autumn Desjardins, Zelda Wiki/Nintendo

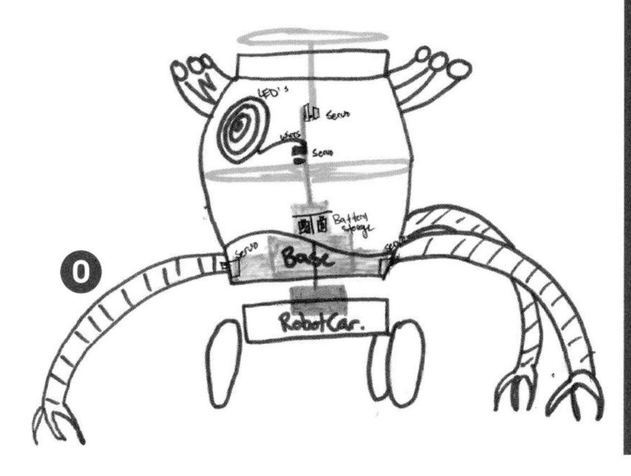

O

Triple Axial

R/C ice skating rig by Ben Eadie, for Anna Kendrick in the Disney+ Christmas movie *Noelle*. The robotic platform drives any direction on the ice, the turntable on top provides faux spin moves, and the tracks — poached from an elastic-band skate trainer — slide to simulate real-life skate gait.

have left and have that display on my remote too, while I am operating the robot.

- Then I take it one step further: I program the transmitter to **disable the robot** at a certain voltage and not allow it to move until I change the battery or charge it!

This simple telemetry setup can save you money in destroyed batteries and also in frustrations of not knowing how much runtime you have left when operating the robot.

What about ***feedback***? Imagine how you could feed sensor data back into your robot's movement — say, move a different way when it detects obstacles or uneven terrain (motor position or load feedback, accelerometer/IMU data), or when it senses lights, sounds, or smart camera inputs, or even a certain tap or nudge from a human companion? Now your prop is becoming a character itself! Literally endless possibilities.

WHOLE NEW LEVELS

OpenTX and EdgeTX are powerful tools that offer advanced features and capabilities for radio control modeling. Understanding telemetry, channel mixing, and how to program the controller are key to unlocking the full potential of your model. My little Mecanum robot will be part of a *Zelda* cosplay by Autumn Desjardins (autumnscosplay.ca), operated with OpenTX wherever she goes (Figure O). With the right tools and knowledge, you'll be able to take your radio control modeling experience to the next level too.

Want More?

- **I share as much as I can on my YouTube channel:** youtube.com/@dreadmakerroberts
- **OpenTX manual:** opentx-3.gitbook.io/manual-for-opentx-2-2
- **EdgeTX manual:** edgetx.gitbook.io/edgetx-user-manual

BEN EADIE is a movie prop maker, practical special effects designer, inventor, nerd, and former aerospace engineer. You can see his work in films like *Star Trek Beyond*, *Predator*, and *Ghostbusters: Afterlife*.

DIY MUSIC

The anthem of our lives

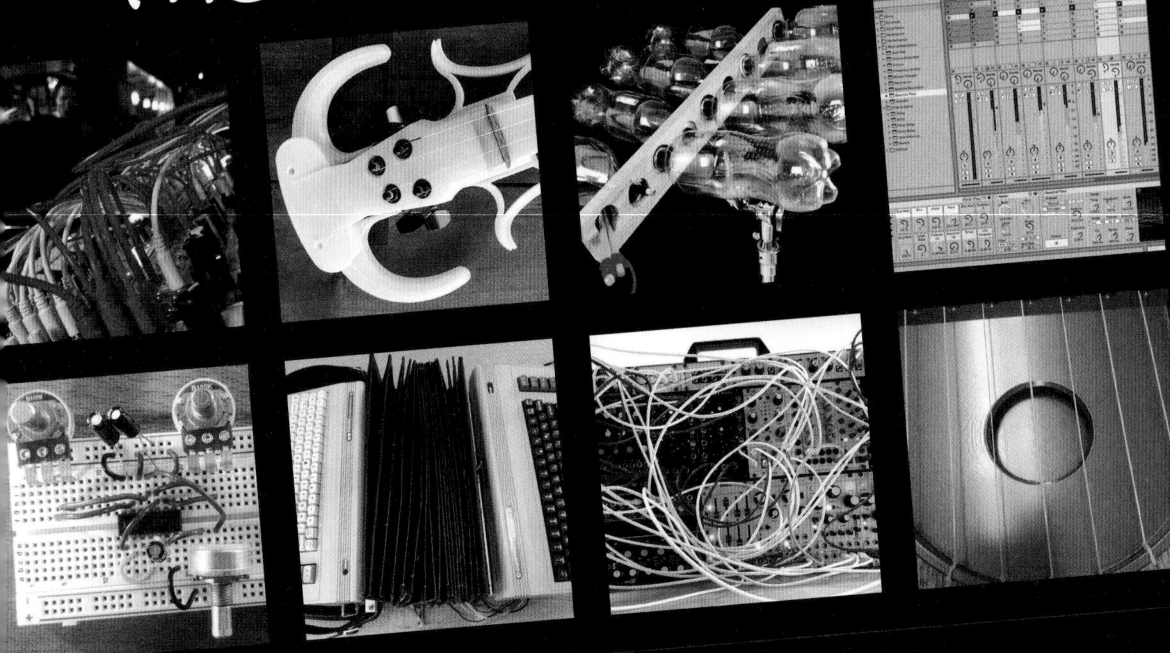

MUSIC. It makes us fall in love. It makes us cry. It is the anthem to revolutions. It's universally beloved yet infinitely diverse.

I listen to music all night and day while I work and play. Live concerts are my obsession and have changed my life in a million ways. Communities built around musical genres and production enrich my existence to no end. Consuming it, creating it — music *is* my life.

When we were discussing themes for this volume of *Make:* and the topic of music came up, my already irrepressible enthusiasm went intergalactic. I've always loved music and surrounded myself with music and music lovers. When I first started getting intro electronics, I needed a "target" for my desire to learn the tools and techniques of this field. I started with simple learn-to-solder kits that made little bleepy-bloopy sounds (in fact they now live in a bin called "Bleep Bloop") such as nootropic design's Audio Hacker and DJ shields for Arduino, as well as Bleep Labs' theremin-like Pico Paso.

One of the most exciting convergences of music and DIY right now is the burgeoning synth DIY (SDIY) scene, which we dive into with modular synthesizers (page 40). Build your own simple synth and learn what makes it squeal with the Mt. Brighton Avalanche Oscillator (page 48). Go low-tech with a great-sounding Soda Bottle Marimba you can build for pennies (page 60). Or 3D print your own speakers and instruments (page 64).

These and many more make up our special DIY Music section — and you'll find more cool sound circuits throughout this issue, like artist Kelly Heaton's beautiful Nightjar project on page 22. From birds to boards to Buchla, there's something for everyone!

So fall in love, cry, start a revolution, attend or perform a live show, build a project from these pages — immerse yourself in the world of music however you like, as we together form the anthem of our lives.

Joe Bauer, obert graham, Jet Kye Chong, Jim Axelsson, Nick Gaydos, Dogbotic Labs, Courtesy of Linus Åkesson, mikolas zuza, Kemi Adejumo, Adobe Stock-Neo

The author performs at v/Oct, a monthly modular event, at The Last Word in Ann Arbor, Michigan.

GETTING STARTED WITH

MODULAR SYNTHESIZERS

A BLAST FROM THE PAST: FROM THE MIGHTY MOOG TO TODAY'S "SDIY" SCENE

Written by Nick Gaydos

You slowly adjust a knob. The panel's LEDs flicker. You flip a switch and feel the low rumble in your chest.

Although the machine before you looks like something straight out of Kennedy Space Center, you're far from Cape Canaveral. Cables, sliders, screens, and jacks all work together to shape sound. The ***modular synthesizer*** before you holds infinite possibilities.

While they may look like they require a degree in rocket science to operate, we hope to demystify how modular synths work, why you might be interested in them, and what you need to launch.

Donald Buchla, one of the two co-inventors of voltage-controlled modular synthesizers, actually worked at NASA before his formative work for the San Francisco Music Tape Center in the early 1960s. Unbeknownst to Buchla, theremin maker Robert Moog was simultaneously working on something similar in New York. Both had independently invented a voltage-controlled modular synthesizer!

Let's back up a bit.

Don Buchla blazed new trails in sound synthesis to co-invent the voltage-controlled modular synthesizer.

Bob Moog and Herb Deutsch collaborated to create the first Moog synthesizer in 1964.

WHAT IS A MODULAR SYNTHESIZER?

A synthesizer is a musical instrument that creates sound electronically. And in a modular synthesizer (Figure **A**), the individual building blocks of a synthesizer are broken out into separate functions.

While many of these building blocks were not new inventions, Moog and Buchla arrived at the same components, which are still used today:

- **Voltage controlled oscillators (VCOs)** provide a sound source
- **Voltage controlled amplifiers (VCAs)** allow the sound to be loud or quiet, on or off
- **Filters** permit only specific frequencies of sound to pass through
- **Envelopes** shape the audio signal over time (attack, decay, sustain, release, or **ADSR**)
- **Noise generators** create random frequencies of white or pink noise
- **Mixers** combine the signals.

Modularity was simply a byproduct of rapid iteration and feedback; adding or changing something without completely reworking it is

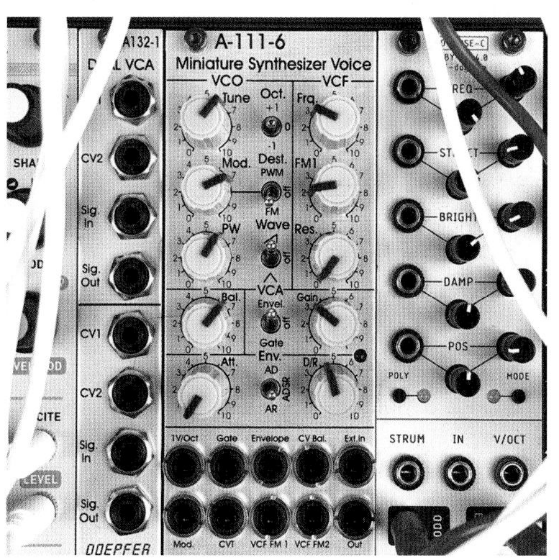

The same building blocks of the earliest modular synths are seen in today's Doepfer A-111-6 module: VCOs, VCAs, filters (VCF), ADSR envelope, and inputs for CV and gate signals.

always much more manageable. With everything separated, the signals must move from module to module. A musician makes those connections using cables to make a ***patch***.

Both inventors used transistors instead of finicky vacuum tubes, but Buchla and Moog's key innovation was that the modules were all ***voltage controlled***. Those voltages could be:

- **Audio signal** — a sound source in the range of human hearing
- **Control voltage signal (CV)** — used to control the different parameters of a module (such as pitch and amplitude)
- **Trigger** or **gate signal** — a pulsed voltage used for timing (on and off, high and low, true or false).

Let's look at each of these in a bit more detail.

AUDIO SIGNALS

If we zoom far out from our little world of synthesis and put our physicist hats on, we see that everything is just a wave of energy. Sound is ripples of energy moving air particles at a frequency between 20Hz and 20kHz (at audio rate, or within the human range of hearing).

Most traditional instruments create sound through vibration. Synths don't have air chambers or strings to vibrate; they rely on the movement of speakers to create sound waves. So synth sounds typically start with an ***oscillator***. This produces a synthetic vibration that becomes sound when sent through a speaker. The speaker moves back and forth with the rise and fall of the oscillator's voltage (positive to negative). Increase the frequency of the oscillation, and the speaker moves back and forth faster, which results in a higher perceived pitch (note).

CONTROL VOLTAGE

Control voltage (CV) is simply a voltage used to change a module's parameters. While most modular synthesizers use control voltage, module designers implement it differently. For example, Bob Moog decided that the frequency on his oscillators would respond to a scale of 1 volt per musical octave. Buchla arrived at 1.2 volts per octave with 100mV per semitone.

That same CV signal could control the amount

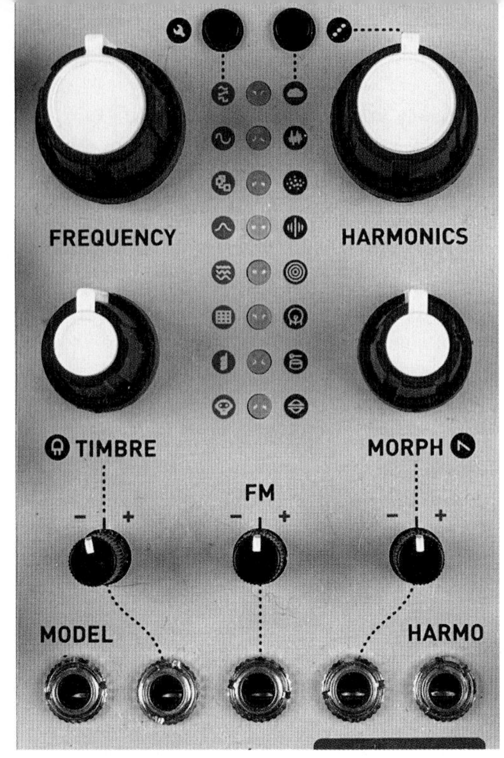

Nick Gaydos, Joe Bauer

Synthesizer sound starts with an oscillator in the audio range, 20Hz–20kHz.

of resonance in a filter or the amplification in a mixer. It's up to the module designer to decide which parameters use CV. Often a parameter has a knob or slider for manual control and a CV input jack that allows voltage to "turn the knob" for you.

Changing parameters over time is called ***modulation***. Let's slow down our audio rate oscillator to under 20Hz. In that case, it becomes a **low-frequency oscillator (LFO)**, a common type of module that's perfect for tempo functions or for modulation.

Turn it down to 1Hz (one cycle of the wave every second) and patch it into another oscillator's CV input for frequency. Once per second, the LFO's voltage will go from low to high, raising the second oscillator's pitch and then dropping it back down. This technique can be used subtly as a vibrato or tremolo effect, or to thrill your neighbors with ambulance sounds.

TRIGGERS AND GATES

Control voltages can be simplified further to control rhythm modules or timing signals. Triggers and gates are simple high/on, low/off voltages that signal the start and stop of an event. Modules often treat gates and triggers

This sequencer is loaded with gate outputs for triggering sounds or samples.

Time for a coffee cup jam session in a local café, with North Coast Modular Collective.

interchangeably.

Gates are significant for the time held at a high voltage. For example, a key pressed on a keyboard creates a gate that signals how long the sound will play.

Triggers are just very short gates. These bursts send signals to other modules to do something, from keeping tempo with a clocked trigger to signaling a drum module to play.

WHY NOW?

When you think of songs with synthesizers, "Jump" by Van Halen, "Sweet Dreams" by Eurythmics, or "Blinding Lights" by The Weeknd might come to mind.

Was any of this music performed on a modular synthesizer? Nope. So what happened?

Early modular synths were notoriously tough to keep in tune. They were expensive, and traveling with something so big and bulky was hard. Many musicians didn't have the patience or time to re-patch between songs.

In 1970, Moog's Minimoog Model D hardwired the paths of the modules, creating a portable synth in a suitcase that was more affordable and easier to understand. Other manufacturers followed suit and soon there were preset sounds, MIDI connectivity, sequencers, samplers, and new types of digital synthesis. The warm analog sounds of modular essentially disappeared.

It wasn't until the 1990s that people began to long for the sound of raw oscillators through analog filters. It was also around this time that Dieter Doepfer introduced his A-100 modular system in Germany. The A-100 was compact, substituting modern 3.5mm (⅛") jacks for chonky old ¼" phono jacks, and other manufacturers quickly adopted its format. The ***Eurorack standard*** gave musicians an easy way to integrate modules from many different companies into their systems.

Eurorack's popularity continues to grow. It is now the most prevalent modular synthesizer format, with hundreds of manufacturers creating modules.

WHY YOU SHOULD TRY MODULAR

Modular synths are one of the few instruments that are both open-ended and tactile. Electronic musicians are drawn to modular because their other favorite open-ended instrument, the computer, isn't nearly as focused. Modular synthesizers can't check email or browse the internet. And electronic musicians certainly love hands-on knob-twiddling. It beats changing one parameter at a time on a screen with a mouse.

While traditional synthesizers are hardwired to follow the signal path the designer created (typically Oscillator→Filter→Envelope-controlled VCA), modular synths can blaze whatever trails your modules allow. With well over 10,000 modules created by the Eurorack community, you'd be hard-pressed not to find one to scratch your itch.

It can also be refreshing to approach the same instrument repeatedly, getting unique results each time. The modular synthesizer is an entirely blank canvas.

One of the author's portable racks features a tangle of patch cables.

WHY YOU SHOULD STAY AWAY

Getting into modular is expensive. It is a small boutique market, and production limitations can lead to FOMO (Fear Of Missing Out) and GAS (Gear Acquisition Syndrome). Your time spent researching modules should be less than the time spent using modules.

The siren of unfilled space will call out to fill your case. Resist temptation by installing blank panels.

And a blank canvas can be both a blessing and a curse. It can be daunting to know where to get started, and it's easy to get lost without approaching the instrument with intention. Experimental production is not always productive experimentation.

THE COMMUNITY

Go to a meetup. Attend a performance. Visit your local synth store if you have one nearby. Ask people for advice on ModWiggler. Everyone was a beginner once, and most are happy to help. Use YouTube's tutorials and talks as a resource. Leave comments and ask questions.

You'll find online communities in every neck of the social media woods, from Instagram to Mastodon to Twitter to Reddit to Facebook to Discord. Hashtags are your friends: #eurorack #synthdiy #modularsynth #sdiy

alvin hill performs live at a North Coast Modular Collective meetup on getting started with modular, at the Ann Arbor District Library in Michigan.

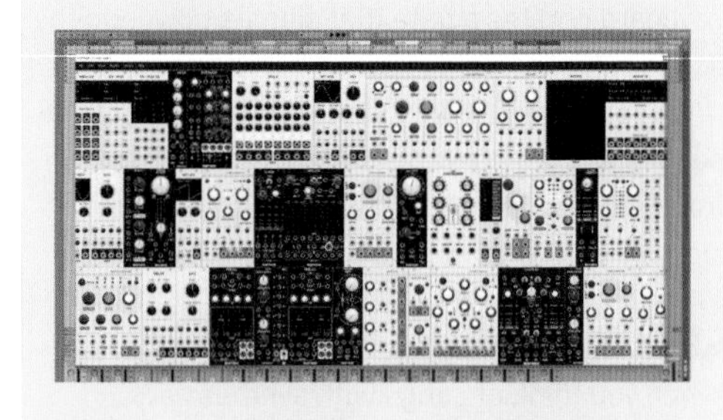

TRY IT OUT VIRTUALLY

VCV Rack is an free open-source virtual Eurorack studio for your computer (see page 59). Use it to become familiar with actual modules and patching techniques before spending money getting physical. There are fantastic YouTube videos from Omri Cohen that will walk you through the concepts of modular synthesis using VCV Rack.

Nick Gaydos, Charles Saadiq, VCV Rack

SETTING UP YOUR FIRST MODULAR SYNTH

THE BASICS

At the very least, you'll need a case and a power supply.

Cases can be as simple as a cardboard box to as fancy as a bespoke piece of furniture. The width of a case is measured in ***hp (horizontal pitch)*** and has a height in ***U (rack units)***. A standard Eurorack module is 3U high.

A clean, reliable power supply is even more important. Some cases will already have a power source installed, but there are plenty of options for cases without power.

A power supply will list amperage for +12V, −12V, and frequently +5V. A good rule of thumb is that the amperage draw of your modules should be well below 75% of your supply's rated output. Modulargrid.net can help determine the power requirements for each voltage.

With the case unplugged, you'll connect the power in your case to each module using a ribbon cable (typically provided with a module). The red stripe denotes the −12V side of the cable and faces down. Remember to match the markings and orientation, as plugging it in backward can destroy a module or your power supply.

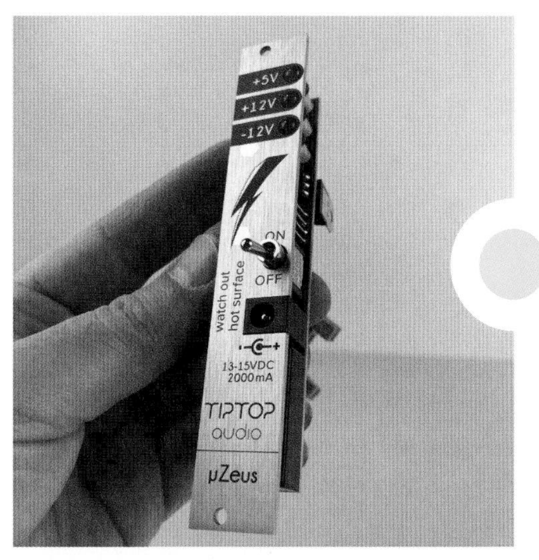

I like the Tiptop power supply for DIY modular cases.

Mutable Instruments Plaits and Braids modules are good starting points, and they're open source designs.

RECOMMENDATIONS:

- **DIY case + Tiptop µZeus** with 3A power supply tiptopaudio.com
- **Tiptop Happy Ending Kit**
- **Intellijel Palette Case** intellijel.com
- **Tiptop Mantis**
- **4ms Pods** 4mscompany.com

STARTING MODULES

You'll want to cram the most functionality into the least space when you're just beginning your journey. Using modulargrid.net to sort and filter through modules will help you pinpoint the perfect fit.

FULL SYNTH VOICES:

A *full voice* module contains a VCO, filter (VCF), envelope, and VCA.

- The **Doepfer A-111-6**, shown on page 41, is a full analog voice with lots of CV control, excellent sound, and a modest price.
- Finding a **Plaits** or **Braids** module from **Mutable Instruments** will give you a flexible digital voice with countless synthesis options. While Mutable has discontinued its modules, its founder Émilie Gillet's generous open-source licensing has allowed makers to continue building upon her legacy.

JACKS OF ALL TRADES:

- **Make Noise Maths** is a flexible analog function generator. It can provide two sets of envelopes and perform logic functions, mix signals, act as

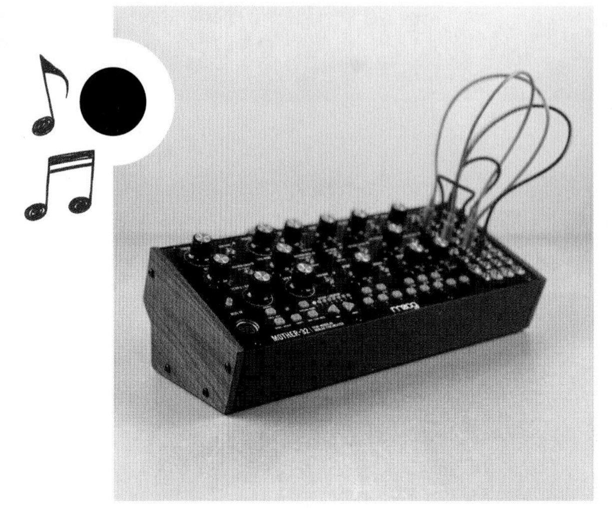

The Moog Mother 32 semi-modular synthesizer packs many module functions into one unit.

A portable(-ish!) modular setup by Joe Bauer includes drum modules, sequencers, and a lot more.

Add an external controller for playability.

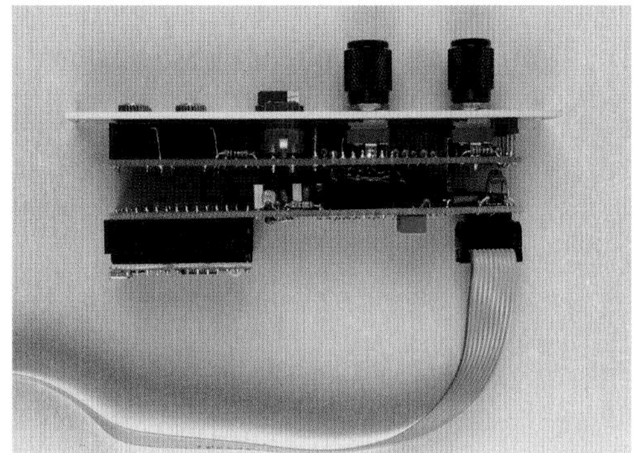

Music Thing Modular makes this Radio Music kit that you can build yourself.

Nick Gaydos, Joe Bauer, Charles Saadiq

a VCA, or even become an oscillator.

- **Expert Sleepers Disting mk4** is a digital Swiss Army knife that can fill almost any gap in a system. With 80+ algorithms, the module can be changed from an oscillator to a reverb to a precision adder. This one requires a bit more practice to master.
- **Pamela's Pro Workout** is an insanely flexible clock source at the heart of many rack systems. Along with its predecessor Pamela's New Workout, it excels at triggering drums, providing Euclidean polyrhythms, and generating interesting modulation signals.

SEMI-MODULAR

Another great path is to start with a *semi-modular synth*. These often have all of the components of a complete modular synthesizer. Still, many connections between modules are "pre-patched," making it easy to get started. Plugging a cable into a jack will bypass the wiring to create your own new sounds.

Semi-modular synths often have a VCO, envelope, VCA, and filter all in one case, including a power supply. They're typically less expensive than buying all the modules individually — all the essentials are already included.

RECOMMENDATIONS:

- **Moog Mother 32 / Mavis**
- **Make Noise 0-Coast**
- **Cre8Audio East Beast / West Pest**
- **Behringer Neutron**
- **Arturia Minibrute 2 / 2s**

WHAT'S NEXT TO EXPLORE?

There's an old adage that you can never have too many VCAs! Some VCAs double as mixers that can combine simple signals to make complex ones. VCAs are also great at taming control voltages using offsets and attenuation. They allow you to "modulate the modulators" to produce unique and exciting sounds.

Filters add character and movement to otherwise identical sounds by removing or accentuating different sound frequencies. Use low-pass filters with envelopes to create a ***low-pass gate (LPG)***. An LPG acts similarly to a VCA but clamps down on sound frequencies instead of amplification.

SNEAKY TIPS

- While modules tend to hold value, you can find great deals by buying used ones.
- Only some things need to go in the case. Save space using external sequencers (like Korg's SQ-1 or Arturia's BeatStep / KeyStep) and external effects (rack mounted or guitar stomp boxes).
- Eurorack audio levels have a 5x–10x greater voltage range than a standard instrument output. However, many stand-alone mixers can handle the hot signals by turning down the gain.

BUILD YOUR OWN DIY MODULES

Putting together your own Eurorack modules from DIY kits or from scratch can save you money and help you learn how they work. (Turn the page to try your first kit build!)

Don't be surprised if you find yourself sketching ideas for your perfect module. The sheer number of Eurorack modules ties directly to makers solving their problems by creating new designs.

The Synth DIY (SDIY) community is incredibly giving and generous with their time. Don't hesitate to reach out if you have questions or face challenges.

NICK GAYDOS is a designer interested in the intersection of human experience and technology, and a decades-long creator of electronic music. He works to foster communities for creative expression as a core organizer of the North Coast Modular Collective and a co-founder of Outside Circle Collective.

PATCH INTO MORE SYNTH DIY RESOURCES

ONLINE SDIY COMMUNITIES

- ModWiggler's Music Tech DIY modwiggler.com/forum/viewforum.php?f=17
- Facebook facebook.com/groups/synthdiy
- Reddit reddit.com/r/synthdiy
- Electro Music DIY Hardware and Software electro-music.com/forum/forum-112.html

SYNTH CIRCUITS IN PRINT

Getting Started in Electronics by Forrest Mims forrestmims.com
Make: Electronics by Charles Platt makershed.com/platt
Make: Analog Synthesizers by Ray Wilson makershed.com/products/make-analog-synthesizers
Electronotes by Bernie Hutchins electronotes.netfirms.com
Multiple books by Thomas Henry lulu.com/search?contributor=Thomas+Henry

GREAT FIRST KITS

- Lodi Mult northcoastmodularcollective.com/modules/lodi-mult-passive-multiple
- Startup by Music Thing Modular thonk.co.uk/shop/startup
- Radio Music by Music Thing Modular thonk.co.uk/shop/radio-music-full-diy-kit

LEARN BY BUILDING

- AI Synthesis Modules aisynthesis.com
- mki × es.EDU DIY System by Erica Synths and Moritz Klein ericasynths.lv/shop/diy-kits-1

KITS, PANELS, PCBs, AND PARTS

- thonk.co.uk
- synthcube.com
- modularaddict.com
- taydaelectronics.com
- mouser.com
- digikey.com

GREAT SDIY YOUTUBE CHANNELS

- Aaron Lanterman @lantertronics
- Kristian Blåsol @kristianblasol
- Look Mum No Computer @lookmumnocomputer
- Moritz Klein @moritzklein0
- Synth DIY Guy @synthdiyguy
- The AudioPhool @theaudiophool
- Under the Big Tree @underthebigtree

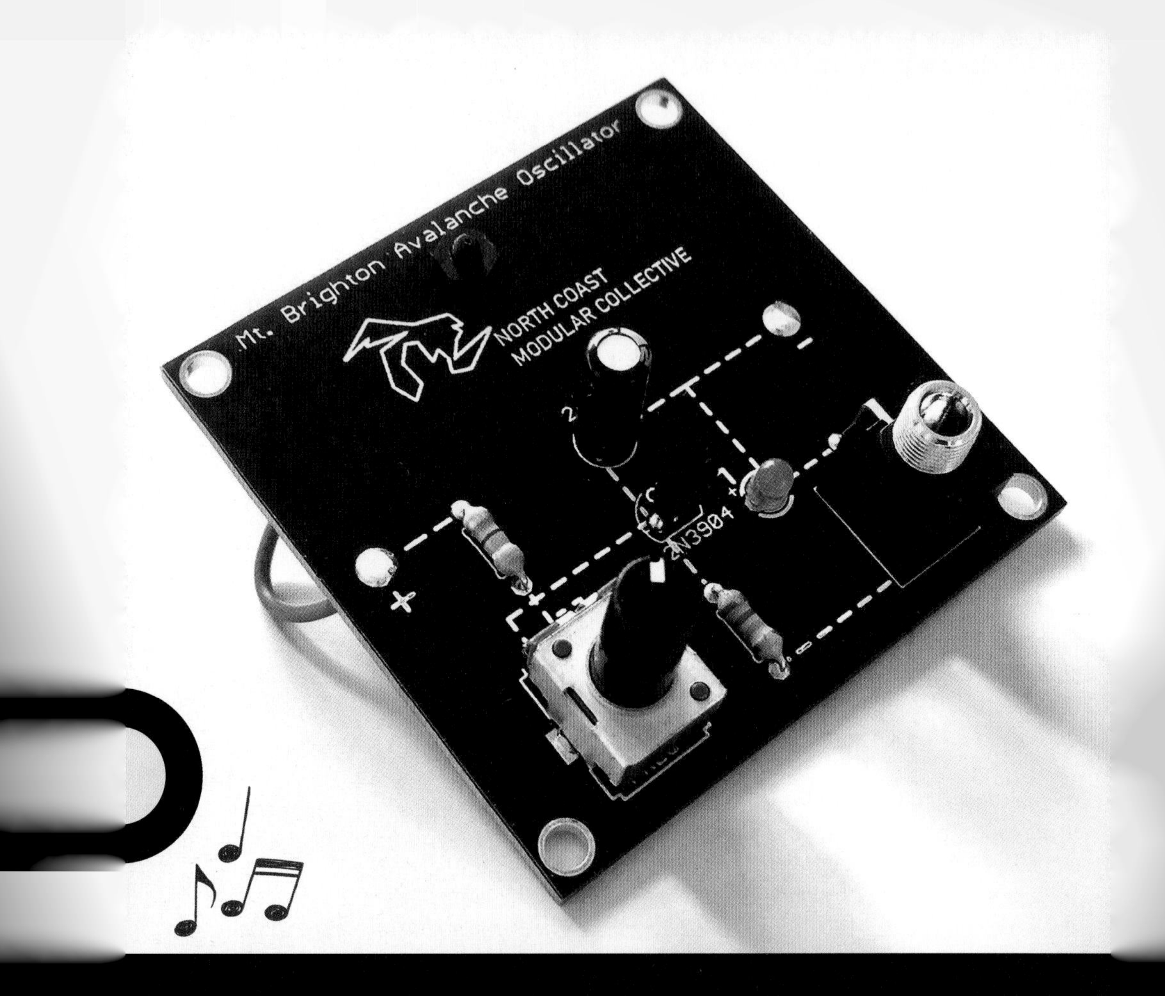

BUILD YOUR FIRST MODULE

Written by Joe Bauer

START ON THE BUNNY HILL WITH THIS DIY AVALANCHE OSCILLATOR!

JOE BAUER is co-founder of the North Coast Modular Collective. Based in Ann Arbor, Michigan, he's been exploring the interactions between information, sound, and visual arts through electronics since the early 1990s, often with instruments he built or coded himself.

TIME REQUIRED: 30–60 Minutes

DIFFICULTY: Easy

COST: $25–$30

MATERIALS

Available as a kit from North Coast Modular Collective (ncmc.link/makeosc) or you can source the parts yourself:

- » **Printed circuit board, Mt. Brighton Avalanche Oscillator v4** from oshpark.com/shared_projects/8SeW5Gm5
- » **Resistors: 330Ω (1) and 22kΩ (1)**
- » **Electrolytic capacitor, 2.2μF** such as Panasonic ECA-1HM2R2I
- » **Transistor, 2N3904**
- » **LED, red, 3mm**
- » **Tall trimmer potentiometer, 9mm, linear 10kΩ** such as B10K from modularaddict.com
- » **Audio jack, 3.5mm (⅛"), Thonkiconn style** such as PJ301M-12 or the new improved PJ518M
- » **Battery, 12V, A23 size**
- » **Battery holder, A23, with wire leads**
- » **Machine screw, nylon, M2.5 x 6mm**
- » **Nut, nylon, M2.5**

TOOLS

- » **Soldering iron**
- » **Solder, non-leaded**
- » **Solder flux (optional)** recommended
- » **Flush diagonal wire cutter**
- » **Phillips screwdriver** to fit M2.5 screw head

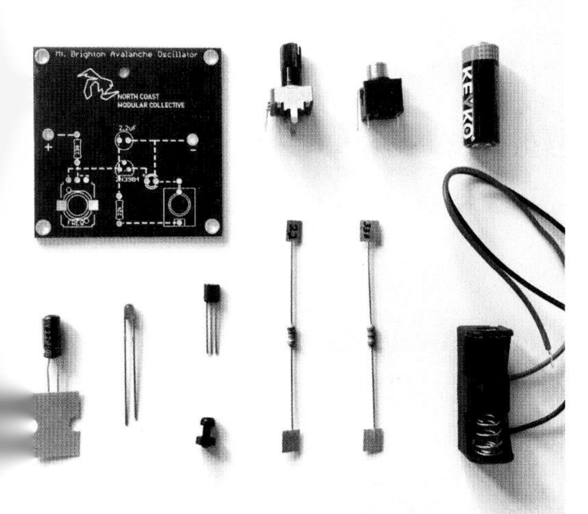

Building your first modular synth is astonishingly easy! The Mt. Brighton Avalanche Oscillator is a community-developed and open-source project designed specifically for learning. Its low part count, low complexity, and low cost is coupled with a transparency in design and function that makes it easier to understand and describe what is happening with the electricity and sound.

When it's complete, you'll have a simple, standalone battery-operated synthesizer, with a single knob for changing the pitch of the sound, into which you can plug headphones or patch other music gear. Assembly takes just 30–60 minutes with basic soldering skills and tools.

WHAT'S IN A NAME?

Mt. Brighton is a ski hill in Michigan that features a "bunny hill" for beginners. And this project takes advantage of a specific trait in transistors called ***avalanche breakdown***: when operating in avalanche mode, a transistor can switch an electrical current rapidly — at an audible rate! This repeated switching is known as ***oscillation***.

North Coast Modular Collective (NCMC) led the design of this project. Our mission is to expand the skill, knowledge, and accessibility of the electronic music community through the development of music, instruments, and shared resources. One of our core values is a sense of place and its connection with the community. As such, NCMC acknowledges that Mt. Brighton currently resides on land ceded through the Treaty of Detroit in 1807 and is the ancestral lands of the Bodwéwadmi (Potawatomi), Mississauga, Anishinabewaki ᐊᓂᔑᓈᐯᐧᐊᑭ, and Peoria nations.

BUILD YOUR AVALANCHE OSCILLATOR MODULE

1. RESISTORS

The project includes two resistors. To "read" a resistor, find the resistor color code (indicated by the bands along their length) online; otherwise, just trust us that orange-orange-brown-gold means 330 ohms (Figure **A** on the following page) and red-red-orange-gold means 22k (22,000) ohms (Figure **B**). Place each resistor as indicated on the PCB silkscreen (Figure **C**) by bending their legs at right angles from the

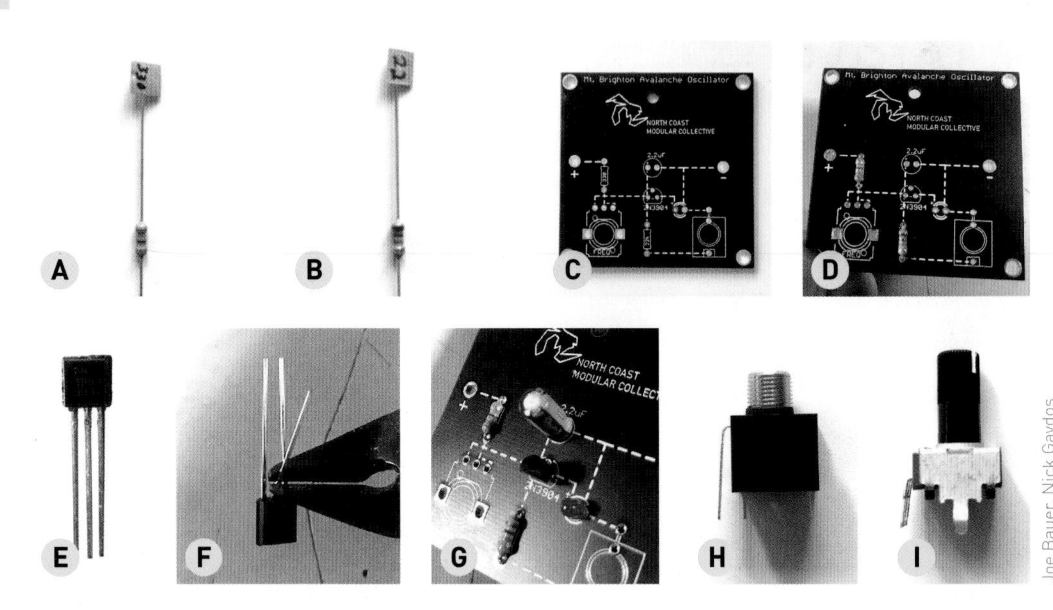

component and pushing them through the holes, aka *through-hole vias* (Figure D). Resistors do not have polarity: place them either way round.

Angle the legs by splaying them slightly outward so that the components are held in place. Solder each leg by heating both leg and pad (the shiny ring around the via) and feeding it solder to create a joint. An ideal joint will be shiny and smooth with a slightly concave surface. (If your joint appears dull or has too much or too little solder, don't worry — you can rework it, just be careful not to heat the components or board excessively with your attempts.)

Now snip the legs off with diagonal cutters to keep things tidy and prevent short circuits.

2. LED

Unlike the resistors, LEDs do have polarity. The longer leg indicates the *anode*, or positive leg. LEDs also typically have a "flat" side, indicating the *cathode*, or negative leg. If you're not certain, use the diode setting of a multimeter to check.

Push the legs through the holes for the LED, marked on the PCB with two concentric circles. Be sure that the long leg is through the + hole and the flat side is opposite it. Solder and snip.

3. TRANSISTOR

The transistor (Figure E) is what creates the oscillation in our circuit, and the oscillation is what creates the sound! Remove the middle (base) pin from the transistor with the diagonal cutters (Figure F), cutting it as short as possible. Push the remaining outer legs through the holes marked 2N3904 on the PCB, with the flat side toward the bottom. Bend, solder, and snip!

4. CAPACITOR

While some capacitors do not have polarity, this 2.2µF electrolytic capacitor does, as indicated by the large minus (–) symbols aligned with the negative leg on the component's case. Find the 2.2µF label on the PCB, ensure the – leg goes through the hole *opposite* the +, make sure it's fairly flush to the PCB, then bend, solder, snip! Your board should look like Figure G.

5. AUDIO JACK

In order to get sound out of our circuit, it's connected to a 3.5mm mono jack. Orient the three pins as indicated on the silkscreen, then solder and trim (Figure H).

6. POTENTIOMETER

The trimmer potentiometer (or "knob"!) lets you change the pitch, or frequency, of the sound by varying the resistance. Because of the force that you'll exert twisting this small pot, it has two heavier legs (Figure I) in addition to its three active pins. These may take a little more solder, but be sure to fill the larger through-hole via and create nice joints so your potentiometer is secure.

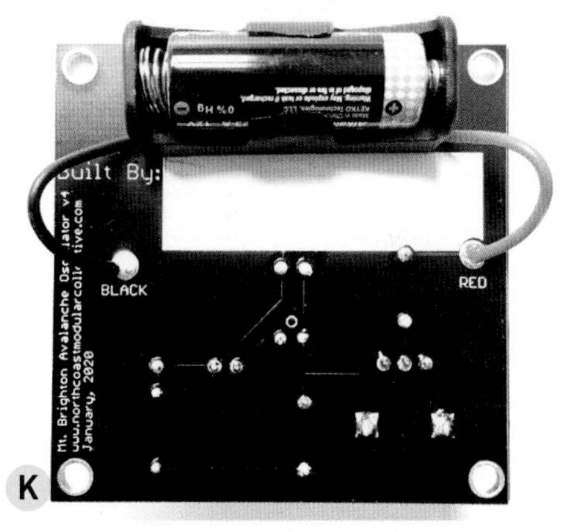

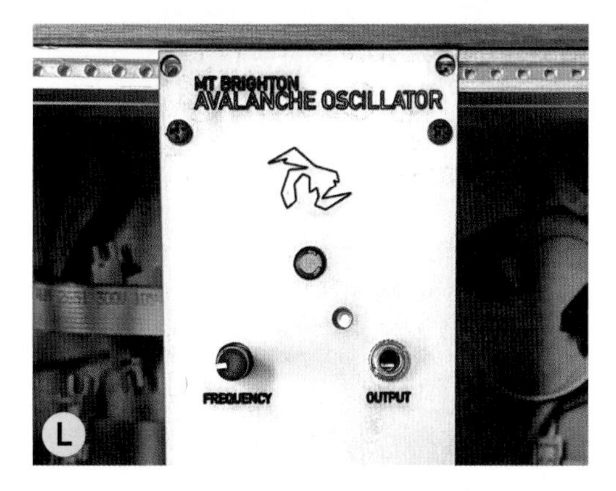

7. BATTERY

Your circuit is almost complete — except it lacks power! In order to not have to deal with voltage conversion, we're using a 12V battery to provide the juice we need. The A23 battery holder is affixed to the PCB with a nylon screw and nut (just finger tight is fine). In order for the holder to lay flat against the PCB, the wires should sit within the indentations on either side of the holder.

Trim the positive (red) and negative (black) wires so they can comfortably reach the RED and BLACK vias on the PCB. Strip about ¼" of insulation from each wire with wire strippers or cutters. Twist the braided wire so the strands are nicely organized together, then put the wires through the holes from the back, so they stick out the front, where you'll solder and trim.

8. FINISHING TOUCHES

Review your board to ensure all solder joints are smooth and shiny, nothing is short-circuiting, and everything is placed as indicated in Figure J. Next, celebrate your success by writing your name on the white portion of the PCB! Add the battery, with the negative side against the spring.

There's no on/off switch, so as soon as the battery completes the circuit (Figure K), we're live! The LED should be lit, and nothing should be smoking — if either of these are not true, remove the battery immediately to start troubleshooting.

KNOB TWIDDLING TIME!

Plug a portable speaker into the mono jack, set the volume level, and tweak the potentiometer knob to your heart's content. That's a ***sawtooth*** or ***ramp wave*** you're hearing; you can drive the pitch from brappingly low to squealingly high!

Congratulations — you just created an electronic instrument! Now that you've mastered the bunny hill, you'll be crushing double black diamonds in no time. What's next?

- Have access to guitar pedals or other audio effects? Try plugging into those. Add reverb and echo and you have instant drone music!
- See the bigger holes at the corners of the PCB? Those can be used for standoffs or mounting it within an enclosure of your own design.
- Want to be able to control when it makes sound and when it doesn't? Splice in an extension cord and solder an arcade button for on/off.
- Synthesizer DIY folks will note that this module uses the same voltage as Eurorack modular synthesizers (see page 40). For advanced SDIYers, it's possible to modify this project to become a Eurorack module (Figure L).

See and hear the Mt. Brighton Avalanche Oscillator in action: instagram.com/p/B9aUXT4hFFB

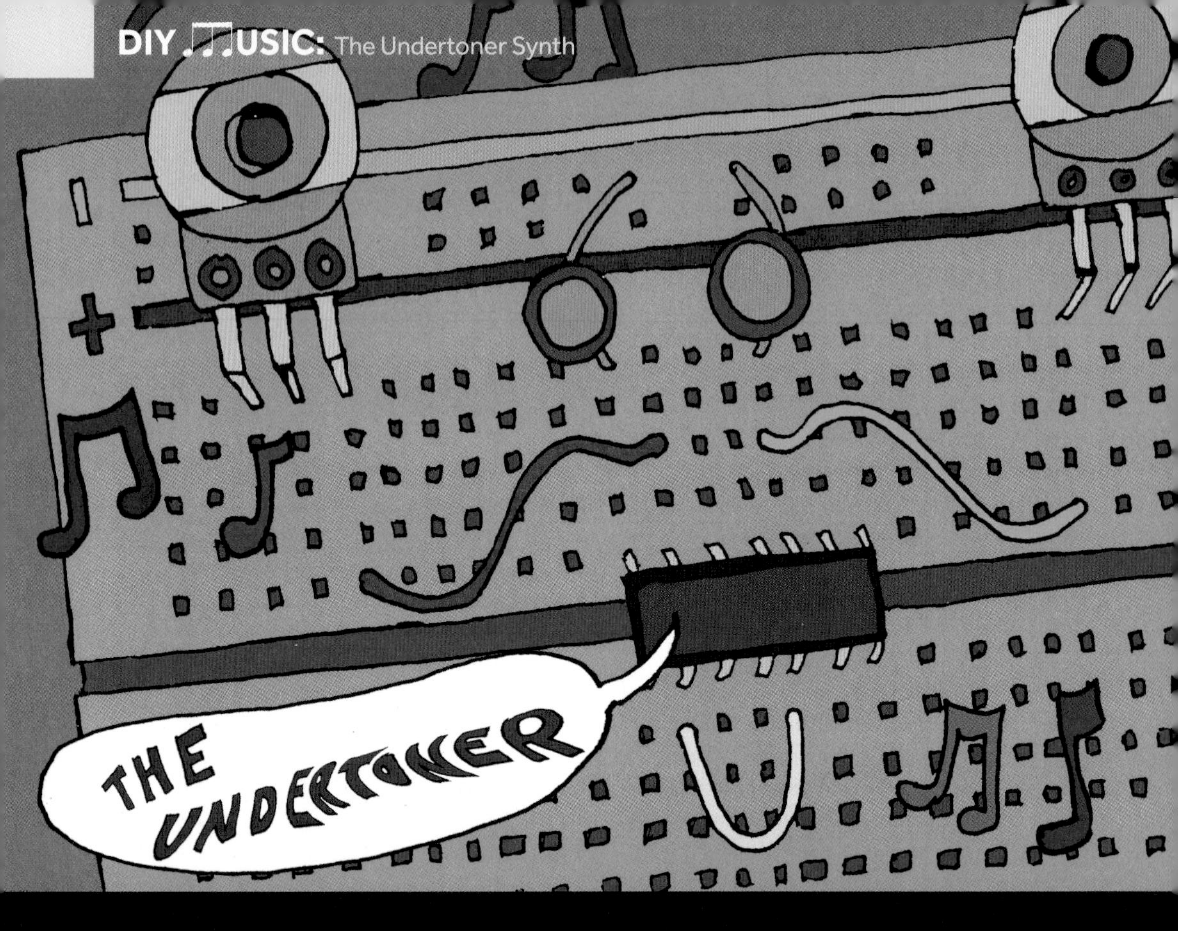

PITCH PERFECT

BUILD THE UNDERTONER — A SIMPLE BUT MYSTERIOUS SYNTHESIZER CIRCUIT THAT ALWAYS PLAYS IN TUNE!

Written by Sean Hallowell and Kirk Pearson, Dogbotic Labs

DOGBOTIC LABS is a San Francisco Bay Area-based audio laboratory and a friendly community where people from all walks of life can experiment and learn together. They offer a beginner's workshop on DIY synthesizers, and whimsical virtual courses on homemade drum machines, video synthesis, experimental photography, wearable electronic instruments, and more. dogbotic.com/labs

TIME REQUIRED: 20–30 Minutes

DIFFICULTY: Easy

COST: $10–$15

MATERIALS

- **CD4093 quad NAND gate integrated circuit (IC) chip**
- **Jumper wires, male-male**
- **Potentiometers (3)** 100kΩ to 500kΩ is a good place to start.
- **Photoresistors (2)** aka light-dependent resistors (LDRs)
- **LEDs** a few to experiment with
- **Resistors** a few in the range of 330Ω to 1kΩ
- **Electrolytic capacitors: 0.1μF (1), 1μF (1), and 10μF (1)** Note that 0.1μF (microfarad) caps are also known as 100nF (nanofarad).
- **Solderless breadboard**
- **9V battery and battery terminal connectors**
- **Alligator clips**
- **Amplifier** Guitar amps or old computer speakers work great!
- **Cable** to plug into your amp

TOOLS

- **Wire cutters/strippers (optional)** to make neat little jumper wires

Every once in a while, along comes a circuit so simple yet so profound that it makes you reconsider the very notion of musical culture.

But more on that in a minute.

The Undertoner is a synthesizer project perfect for beginners looking to dip their toes into the world of electronic music from scratch. This irresistibly musical circuit is built from a single integrated circuit (IC) chip — by name, a *quad two-input NAND gate* (catchy!) — that was not primarily intended for artistic use. However, thanks to an interesting property of said NAND gates, we can conjure a consonant array of musical pitches from mere wafers of silicon. Somehow, the brainless handful of wires you're about to construct *knows how to play in tune!*

But first things first. We'll start by explaining how to put together this simple circuit, with no presumed electronics experience. We'll then explain how this marvelous musico-electronic quirk works, and even walk you through a variation of it in which a flickering light composes melodies before your very ears! Just in time for your next candlelit soirée.

NOTE: Never built an electronic circuit before? No problem! Check out this shopping cart (dogbotic.com/undertoner) with all the requisite parts.

This project is adapted from the upcoming book ***Make: Electronic Music from Scratch***, available for pre-order at the Maker Shed (makershed.com) and other booksellers.

ABOUT THE NAND GATE CHIP

Let's meet your 4093 integrated circuit. This lil' fella is remarkably simple. It's a ***NAND gate:*** a doodad with two inputs and one output. If you connect both inputs to 0 volts of potential energy (assuming you've powered the chip correctly), the output will spit out current. If you connect both inputs to the positive voltage powering your circuit, however, the output will abruptly shut off current, just like that one time you forgot to pay the water bill.

Your computer probably has millions of NAND gates in it — it's an essential ingredient for delicious digital logic. Your 4093 looks a bit different than the NAND gates inside your computer, though; those are miniaturized to make space for everything else, whereas this one's in a ***dual-inline package (DIP)*** form factor. That's fancy-speak for "big enough to be used by human fingers."

Also of note, your 4093 has not one but four of these magical gizmos inside it. That said, what we're about to do will surely make you appreciate NAND gates a lot more than you did already — likely a low bar for success.

MAKE YOUR UNDERTONER SYNTHESIZER

1. POWER THE CHIP

To begin, put the 4093 IC somewhere on your breadboard. Anywhere on the central trench will do. Make sure the semicircle on the IC is facing to the left, as shown in Figure A on the following page.

As oriented here, IC pins are numbered from

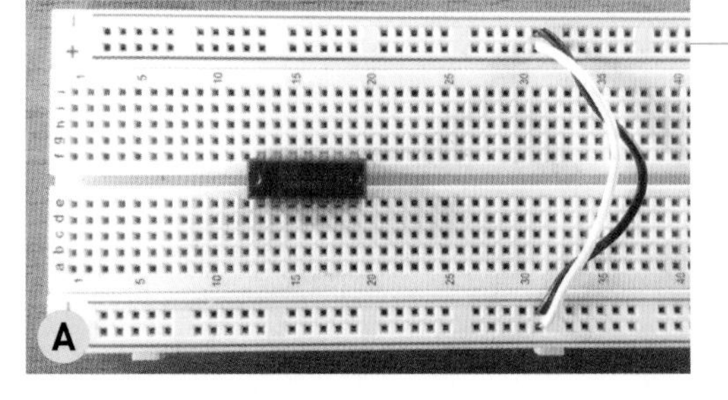

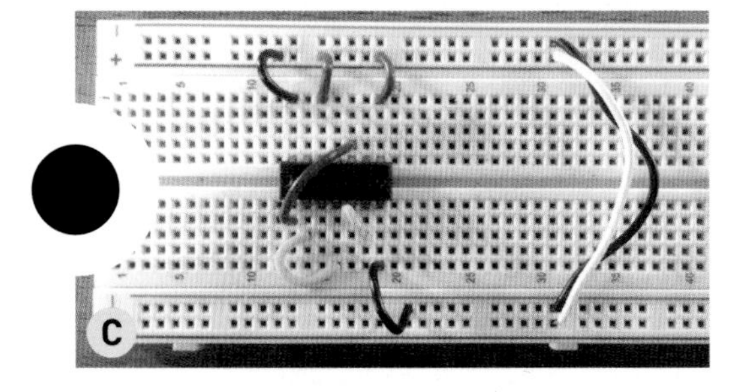

Dogbotic Labs, Maisy Byerly (maisybyerly.com)

the bottom left, counterclockwise around the right side to the upper left. So on a 14-pin IC like your 4093, pin 1 is bottom left, pin 7 is bottom right, pin 8 is top right, and pin 14 is top left (with everything you'd expect in between).

In order to properly power your NAND gates, you'll need to connect them to a power source. 9V will do. This will eventually be piped into your breadboard power and ground "rails," the top and bottom rows marked positive (+) and negative (–), It's not necessary to use these, but they are convenient.

To make sure these rails are sharing electrons, connect those on the top of the breadboard to those on the bottom (positive to positive, and negative to negative). It should look something like the right side of Figure A.

Next, connect pins 14, 12, and 8 to a positive voltage rail (+), and pin 7 to a ground rail (–), as shown in Figure B.

2. CONNECT THE LOGIC PINS

Now make the internal chip connections, as shown in Figure C:

- Pin 1 goes to pin 10
- Pin 2 goes to pin 4
- Pin 5 goes to pin 6

3. ADD THE CAPACITORS

Take a 1µF capacitor and place the long leg (aka the *anode*, or positive terminal) into the row shared by pin 3, and the short leg (the *cathode*, or negative terminal) to the row shared by pin 6 (Figure D).

Our next capacitor is of the 0.1µF varietal. Put the positive leg into the row shared by pin 9, and the negative leg to ground (Figure E).

Our final capacitor is 10µF. It sits between pin 13 (long leg) and ground (short leg) (Figure F).

4. ADD THE POTENTIOMETERS

Now for the pots. You'll attach your first

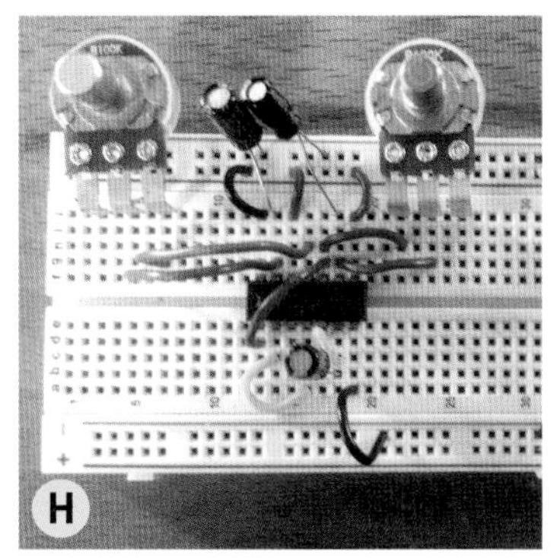

potentiometer between pins 11 and 13 (Figure G). One of the wires will connect to the middle pin (or "nose") of the potentiometer. The other wire can connect to either of the side pins (or "cheeks"). Which of the cheeks you connect simply affects the effect of turning the pot "up" or "down" one way or the other.

The second potentiometer is connected between pins 9 and 10 (Figure H).

The final potentiometer is the trickiest. You'll need to connect one cheek of the potentiometer to pin 6, and the nose to pin 11 (Figure I).

Take a deep breath, because we're done. Now for the moment of truth!

5. PLUG IN AND PLAY!

Grab that cable that's plugged into your amplifier and put two gator clips on the connector plug: one attached to the tip and one attached to the sleeve of the connector (Figure J). The connector shown here has a *tip*, *ring*, and *sleeve* — the ring being the middle portion. (This allows two channels — stereo — to be transmitted on one cable, though our synth is as of yet mono).

Take the other end of the tip gator clip and attach it to a wire going to pin 4 of your 4093. Take the other end of the sleeve gator clip and attach it to ground, or battery minus, also through a wire. With the battery connected, turn up your amplifier volume *slowly*.

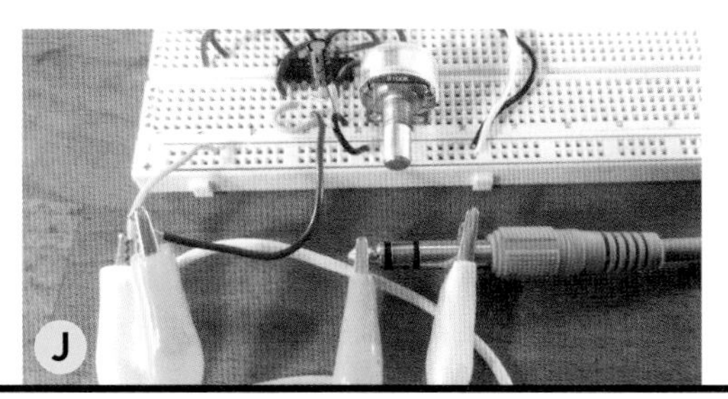

You should hear tasty square waves turning on and off at a steady rate that's controllable by potentiometer 1, the first one you hooked up.

NOTE: Our fourth (unused) NAND gate seemed a little sad over there all by its lonesome, so we turned it into an oscillator to add a rhythmic effect to our already-impressive circuit! If you want to hear the collection of pitches without this gating effect, replace the connection at pin 11 going to the nose of the third potentiometer with a connection to ground.

Now try twisting potentiometer 2 — depending on which way you turn it, you should hear a gradually upward or gradually downward gliding pitch that periodically resets (i.e. jumps back down/up) before starting its ascent/descent again. Adjust it so that you get a relatively high pitch (think "squealing" rather than "bellowing").

Now head over to potentiometer 3 and gradually start turning that one. You should hear [drumroll please] — a series of pitches! One after

the next, merrily up and down, relating to one another in a mysteriously harmonious way. Your circuit is playing not just any pitches but musical notes!

How mysterious indeed! What could be animating such a device? Well, we'll tell you... Right after these hot tips!

VARIATION 1: THE ATARI PUNK CANDLE

If you're the experimental type, now would be a good time to swap out pot 3 for an LDR, aka light-dependent resistor. This way, you can control the transitions between notes with changing light intensity! Grab a candle, enter a dark room, and see if you can find the sweet spot where the candle's flickering makes your Undertoner arpeggiate like a late 70s disco hit.

NOTE: If you're a budding synthesist, you might recognize this scale (and project namesake) from another classic beginner project — the so-called "stepped tone generator" or Atari Punk Console — see page 82 for a new version!

VARIATION 2: LIKE A VACTROLLING STONE

Instead of using a NAND gate oscillator to gate our synth, what if we could use it to arpeggiate our scale? This variation is quite similar to the Atari Punk Candle, but with a twist! Rather than using the oscillator on pins 11, 12, and 13 to gate the synth, you can make it flash an LED on and off. Then position the LED right next to the photoresistor, so the maximal amount of light is hitting the photoresistor during the "on" portion of the cycle. Now, as your light switches off and on, you'll hear the circuit quickly cycle through the notes in the undertone series.

This goofy strobe light-controlled modulation source is called a ***vactrol*** and, believe it or not, these are *everywhere*. If you've ever plugged a MIDI cable between two musical instruments, all you've really done is hooked up several pins to a bunch of tiny light sources that flicker on and off. Because the information is communicated here optically — in *photons* and not *electrons* — the communicating systems are entirely electrically distinct. This way, you can have your MIDI cable connect any two devices without having to worry about blowing them up with the wrong voltage.

HOW THE HECK IT WORKS

The Undertoner is, in a word, baffling. A few logic gates somehow give us a collection of pitches that, no matter how hard they try, simply cannot play "out of tune." How in the world do a bunch of subatomic particles understand a musical phenomenon with such deep cultural roots?

Let's break this circuit down to make sense of it. First, let's talk about how to make a NAND gate oscillate. In talking about NAND gates, it helps to make what computer scientists call a ***truth table.*** It shows us what the output of our NAND gate will be when our inputs are "on" (represented by a 1) or "off" (by a 0):

NAND Input 1	NAND Input 2	Output
0	0	1
0	1	1
1	0	1
1	1	0

As you can see from this table, if we take one of the inputs to our NAND gate and "tie it high" (i.e. connect it to the positive battery terminal), we effectively have an ***inverter*** at the other input (a logic 0 gets a 1, and a 1 a 0).

Now imagine — if we take that inverting input of the NAND gate and connect it to the output, we create a ***feedback loop.*** By connecting a capacitor to the so-called "feedback input" of the NAND gate we add a receptacle for electrons. Specifically, this cap serves as a receptacle that delays the amount of time it takes for voltage at the output to be registered at the input. This is significant because the input waits for the voltage at its pin to reach a certain threshold before defining its state as 1 or 0. As it fills and empties, the capacitor sets the duration of this process. A bigger capacitor means more time between cycles, i.e. a slower frequency, and thus a lower pitch.

Finally, by connecting a potentiometer between the output and input, we can control this frequency with resistance, slowing down or speeding up the amount of time it takes for the capacitor to fill up. (Think of the capacitor as a bucket, and the resistor as a hose. Bigger hose/smaller bucket, faster fill; and vice versa). In other words, the potentiometer gives us variable pitch control. And you said logic was boring!

The NAND gate we've hooked up is outputting a pulsing alternation of 9V and 0V in a pattern that one might describe as its "jack-o-lantern-core" (OK, maybe that's just us). The technical term for this is a ***square wave*** which is accurate, though not as memorable. Have a look at Figure **K**.

Maisy Byerly (maisybyerly.com)

Consider the square wave that says "50% duty cycle." This NAND gate's output consists of equal, alternating "on" and "off" moments (the jack-o-lantern has regularly spaced teeth). There's nothing written in the Laws of Synthesis that says, however, that our "on" and "off" moments have to be equal. In fact, we can vary the ***pulse width*** (how long the "on" portion is) with one of the potentiometers in your circuit. Look at what happens when we vary the pulse width of our square wave to 75% or 25% of the duty cycle.

ENTER THE UNDERTONES

Take a second to consider what might happen if we make the pulse width *longer* than the actual period of the frequency. Now the pulse width control acts also as a signal blocker. As we increase the pulse width past the period of the oscillator frequency, we start blocking every *second* oscillation. Turns out, blocking every other "on" moment of the wave effectively *halves the original frequency*. This is known in musical circles as "playing the same note an octave lower."

As we continue to lengthen our pulse width, every time it passes a new integer threshold, we'll hear a new fractional subdivision of our original frequency. In this way we'll progress from frequency x to $x/2$, $x/3$, $x/4$, $x/5$, and so on.

We will call this collection of pitches derived by dividing a whole number by a series of integers the ***undertone*** series. That's to say, whereas in nature a fundamental tone generates an ***overtone*** series, containing itself as well as all faster (i.e. "higher") frequencies corresponding to integer *multiples*, in electronics the opposite happens. So instead of octave up, fifth up, fourth up, major third up, and so on, on our breadboard we get octave down, fifth down from there, fourth down from there, major third down from there, and so on.

Thus this Undertoner circuit mirrors the natural overtone series by dividing our input frequencies by whole numbers! It's a lovely reminder of how the principles of music apply to the world of electronics as much as they do the world of mechanics.

INTONATIONAL AFFAIRS

Now, about that very notion of musical culture. You might find it hard to play along with this circuit on other instruments. Since the 1700s, the Western European compositional tradition (think piano, guitar, and so on) has largely bypassed this mathematically derived tuning system, known as ***pure intonation*** or ***just intonation***, in favor of something called ***equal temperament.*** TL;DR, this latter system makes it easier for different instruments to play together in tune.

And yet, while the Undertoner, unlike a piano or guitar, is a "synthetic" instrument, the scales it plays are mathematically derived all the same. And when you remember that, unlike the Undertoner, those acoustic instruments have been *unnaturally* adapted to produce a different, more "artificial" scale, it makes you wonder who the real "synthesizer" is after all!

Ableton, Elf Audio, VCV Rack, Adobe Stock-Neo

BILL VAN LOO is a creative technologist, maker, educator, and musician from Ypsilanti, Michigan. A co-founder of the North Coast Modular Collective, he uses mobile apps, software, and hardware to produce dub techno, instrumental hip-hop, and ambient music.

FIRST TRACKS WITH ELECTRONIC MUSIC SOFTWARE

Written by Bill Van Loo

SAVE SIGNIFICANTLY WITH SOFTWARE SAMPLERS AND SYNTHS!

Software is an affordable way to start making electronic music. Let's look at a short list of popular titles and how you might use them.

But first, think about what's fun for you. Are you inspired to craft your own sounds and textures? Create patterns or small phrases of music? Produce full songs? Sometimes with electronic musicianship there's a strong emphasis on churning out finished tracks, when it's fine to just play for the joy of playing! I approach music apps like a sketch pad for fun ideas, then take those ideas into other software or hardware I use.

1 KOALA SAMPLER

iOS and Android, $5 koalasampler.com

Create your own sounds by recording from the microphone or use samples made in another program, then combine them with 16 different effects. To make short patterns or phrases, you can load a whole bank of samples and easily create beats with Koala's grid-style sequencer.

I use Koala to creatively remix stuff I've made elsewhere. I'll take samples from a studio session, play with the Koala effects, and record to an audio file. Then I can post it to SoundCloud, or pull it back into Koala as a sample and play more, or into Ableton Live as starting blocks for a new song.

2 ABLETON NOTE

iOS only, $6 ableton.com/en/note

Ableton's newest product, Note, has some of the features of their flagship Ableton Live, but in a really simple mobile app flow. Perfect if you're interested in making loops or phrases of music.

On the main screen is a grid of *clips* — little bits of note information or sounds — which you can create by writing them out or by recording them. When tapped, clips will play back in sync with each other, and loop at a specific tempo; e.g. drum sounds, bass, and sound FX. Clips can be samples or synthesizers, similar to instruments on Live but not as flexible and fully featured. You're kind of locked into this little set of watercolor paints, but within that, you can make really interesting stuff. Record to audio files, or transfer your Note session over their cloud to Ableton Live 11 to develop your ideas further.

3 ABLETON LIVE

MacOS and Windows, $99+ ableton.com/en/live
If you're looking to make full songs, this is a great option; everything that comes out of my studio runs through Ableton Live at some point. It offers three versions of digital audio workstation (DAW) software, with the largest, Suite, having some bonkers levels of control and customization.

The main view is a clip-based window, like Note, except you have a multitude of sound source or instrument options — sampler, software synthesizer, MIDI arranger (for external communication), or plugins like Moog's emulated Minimoog synthesizer.

It's fun to sequence a software synth and write automation to control parameters, but it's even more fun to "play" Ableton Live with a MIDI controller. These come in all shapes and sizes: piano keys, knobs and faders, pads and buttons with lights. Map these physical controls to the software controls, and you can step away from the mouse and keyboard and control things with more intuitive expressive interfaces.

4 VCV RACK

MacOS, Windows, and Linux, free vcvrack.com
Modular synthesis has exploded in popularity (see page 40), but it can be expensive. VCV Rack, the open source virtual modular synthesizer, lets you get started without spending a dime! Like the real thing, you place modules into a rack and connect them to create your own customized instrument. Each module has a narrow set of features — an oscillator, a sequencer, and so on — but you can use any number of modules, all free!

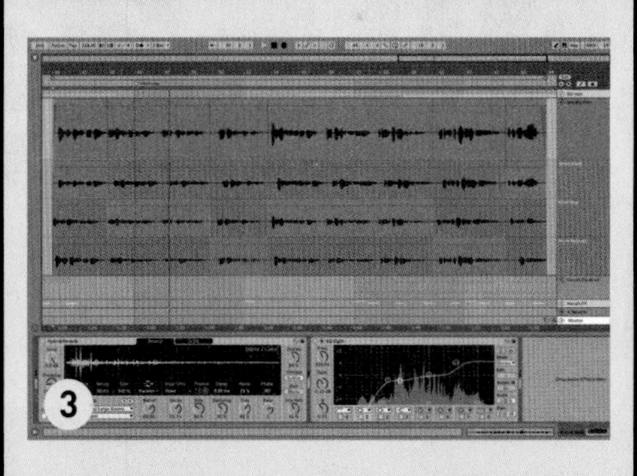

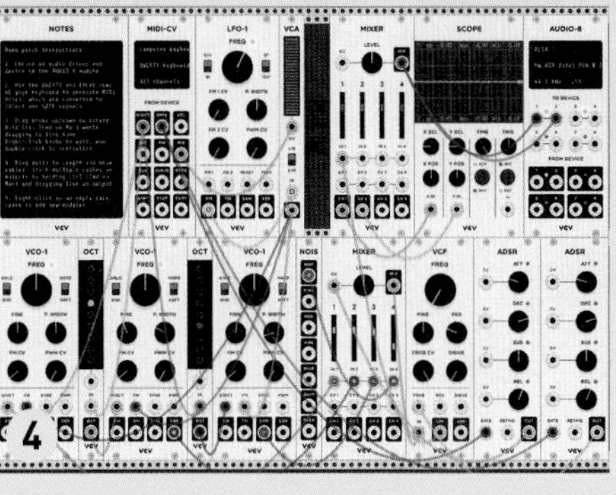

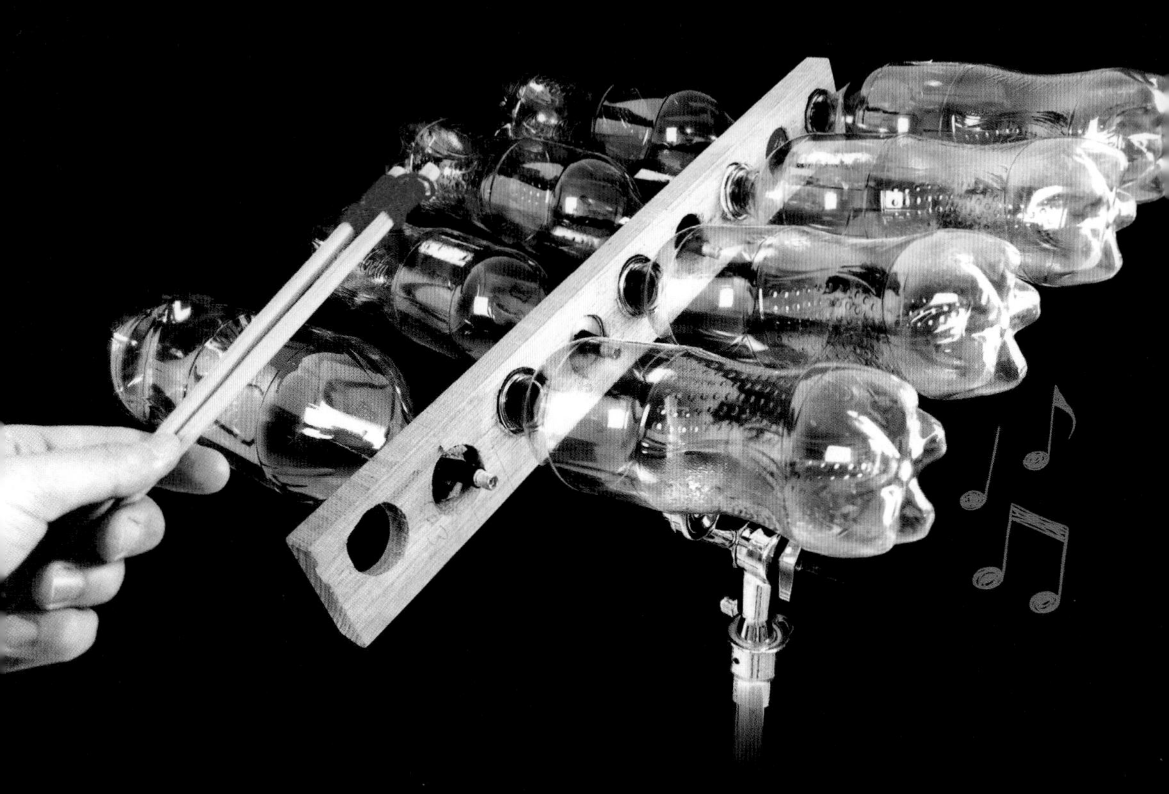

AIR BOTTLE MARIMBA

FUN TO PLAY AND EASY TO MAKE, IT'S GOT A WONDERFULLY PURE TONE

Written and photographed by Jet Kye Chong

Adobe Stock-Neo

JET KYE CHONG is a percussionist and composer from Perth, Western Australia, whose love for making music is rivaled only by his love for making things that make music.

TIME REQUIRED: 1–2 Hours

DIFFICULTY: Easy

COST: $0–$10

MATERIALS

- **Plastic Coca-Cola bottles, labels removed (8)** 600ml/20oz bottles are a good size to get started.
- **Bicycle tire valves, cut from old inner tubes (8)** or car tire valves if you're buying them new, e.g. TR414 tubeless valves
- **Chopsticks**
- **Tape**
- **Scrap wooden plank, e.g. 1×4**
- **Yarn (optional)**

TOOLS

- **Drill and drill bits**
- **Hole saw, 30mm or 1 3/16"**
- **Bicycle pump**
- **Tuner** There are many free tuner phone apps available.

Musical instruments don't have to be expensive and difficult to play. This Bottle Marimba is made of pressurized plastic bottles that make a surprisingly pure, bell-like chime when struck, akin to the sound of a hang drum, and you can tune it to whatever scale you like. It can be built easily entirely from recycled materials and can be played by anyone. It's also a perfect project to build with kids under supervision, encouraging creativity both in building with recycled materials and improvising music.

MAKE YOUR MARIMBA

Before you start, you might like to watch my very short assembly video at youtu.be/kolC_MAB3MU. But it's a very easy build!

1. Drill a hole in each bottle cap slightly smaller than the diameter of the bicycle valves you're using (Figure A).
2. Push the bicycle valve through the cap from the bottom, so that the inlet protrudes through the top. Ensure that there's an airtight seal between the valve and the hole. You may consider deburring the edge of the hole and adding some glue, although I found that a push fit works just fine (Figure B).
3. Screw the cap tightly onto the bottle (Figure C).

4. Slowly pressurize the bottle with the bicycle pump while regularly tapping on the side of the bottle with a chopstick to produce a tone (Figure D). Higher pressure produces higher pitches. Release pressure to lower the pitch.

   Tune the bottle to your desired pitch — some recommendations are included on page 63. Be careful not to over-pressurize!
5. Wrap tape around the bottle caps to get a tight fit into the mounting holes later. I found that

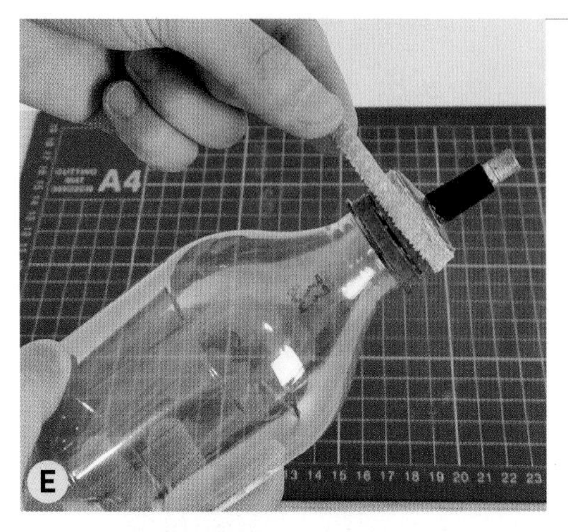

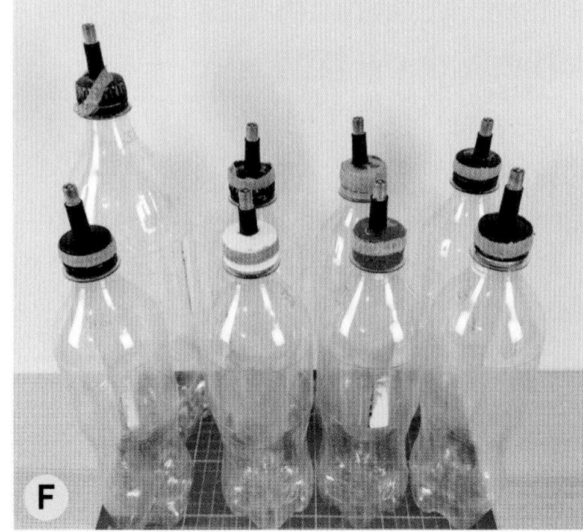

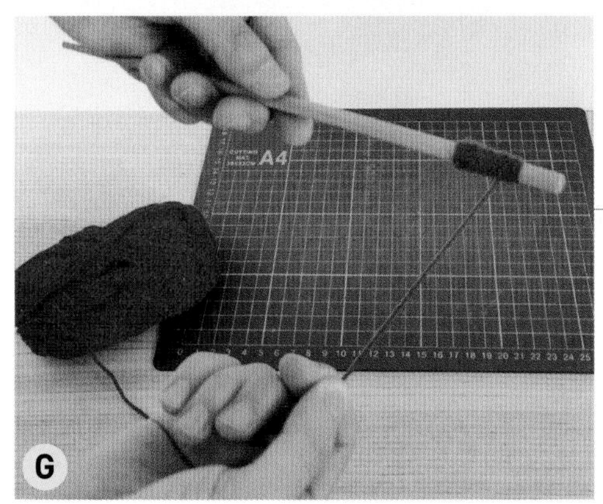

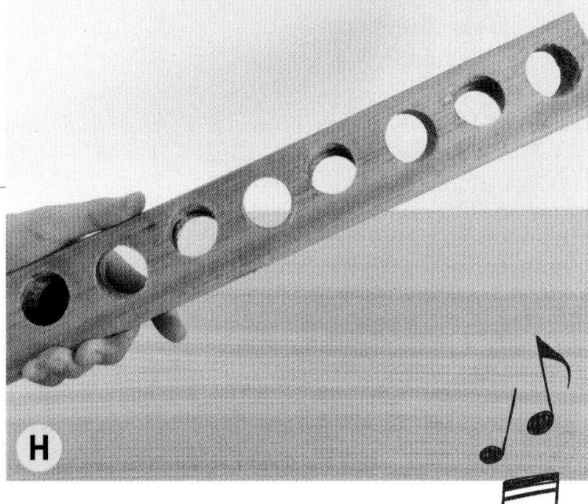

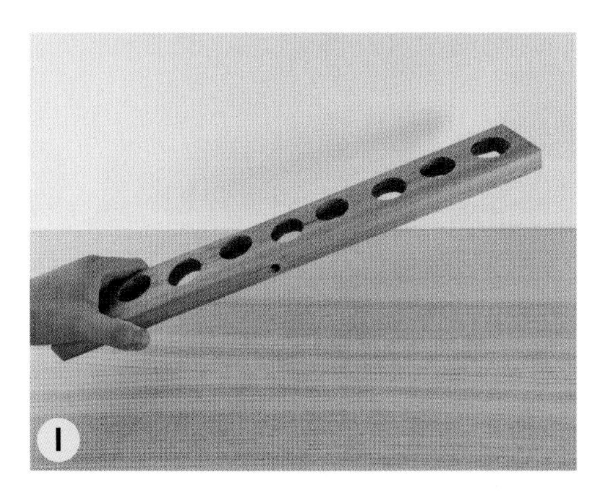

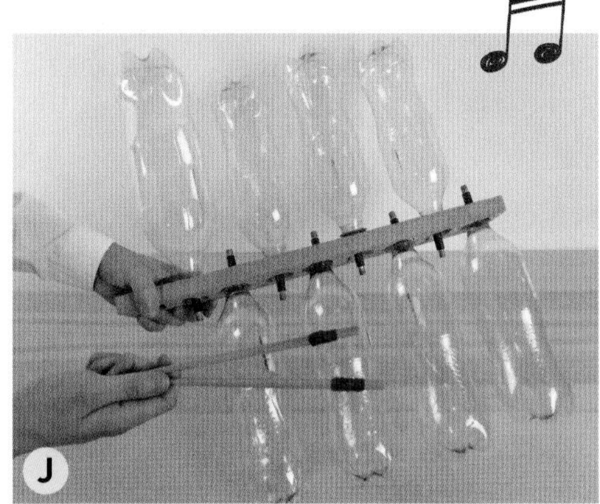

wrapping from right to left (Figure E) helps the tape stay on when screwing the bottles in.

6. Repeat for all of your bottles and bicycle valves (Figure F).
7. Test the bottles by hitting them with the fatter ends of a pair of chopsticks, striking the sides of the bottles near the base. For a more mellow sound, wrap and knot some yarn around the chopsticks (Figure G).
8. To mount your bottles, drill 30mm ($1\frac{3}{16}$") diameter holes into the wooden plank, spaced at least 5cm, or 2" apart (Figure H).

I also drilled a hole in the edge (Figure I) to rest the instrument on a cymbal stand or stake.

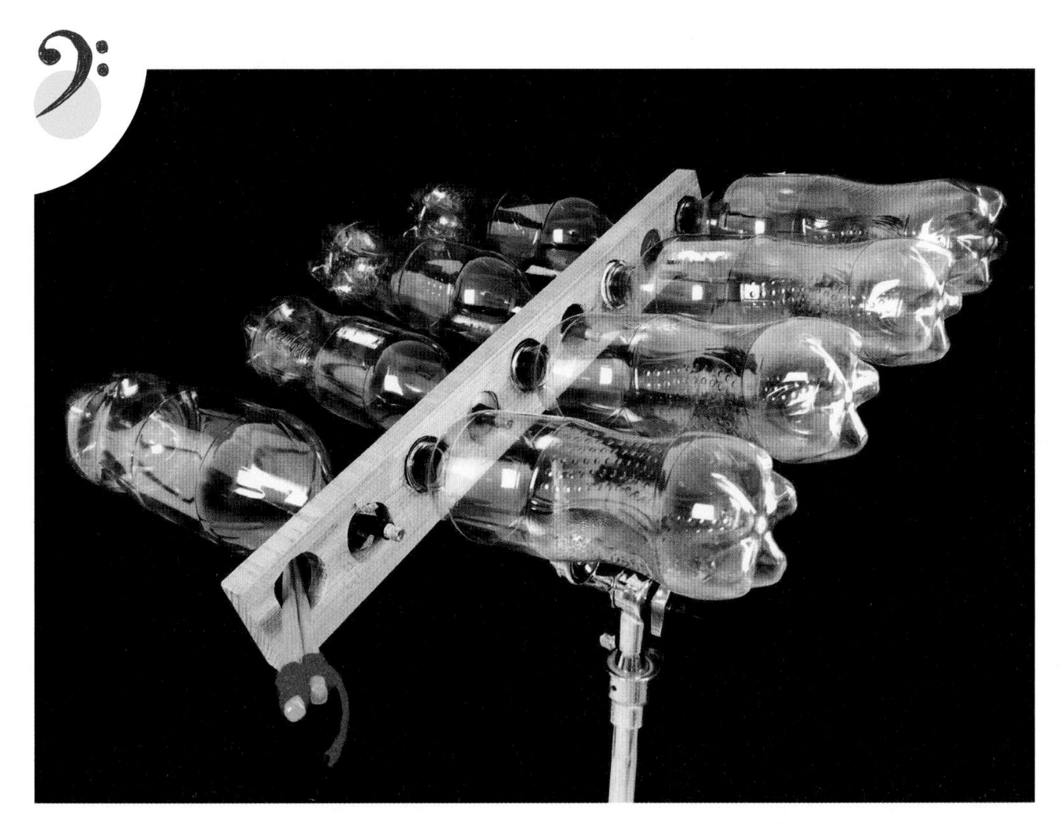

Alternatively, you can tape the plank to a stool or benchtop.

9. Insert the tuned bottles into the holes, alternating sides (Figure J), and you're done!

STRIKE A MELODY

Play your Bottle Marimba by striking the sides of the bottles with the chopsticks. Striking the bottles near the middle tends to give more tone, whereas striking higher up, or on the feet of the bottle, tends to produce more of a "click."

There are many variations to the build that you can explore:

» **Bottle selection:** I've tried many types of soft-drink bottles and found 600ml/20oz Coca-Cola bottles to work well, but larger and smaller bottles work well for lower and higher pitches. Experiment with whatever recycled bottles you have on hand!

» **Valves:** You can often get scrap bicycle inner tubes from bicycle shops for free or cheap. Alternatively, if you buy new valves I recommend buying car tire valves instead, as their rubber coating is thicker and easier to seal against the bottle cap.

» **Tuning:** Here are a few tuning ideas. I recommend starting with the pentatonic tuning which will always sound pleasant. You can hear my G pentatonic tuning in this video: youtu.be/RpjaNeT_mTA.

- **Pentatonic:** G A B D E G' A' B'
- **Aeolian pentatonic:** G A B♭ D E♭ G' A' B♭'
- **Major:** G A B C D E F♯ G'

» **Mounting:** Using a wooden plank is just one way of mounting the bottles. For example, you could drill holes into the bottom of a mixing bowl instead for a circular arrangement.

My friends over at Kaboom Percussion took this instrument to the extreme, building over three octaves of chromatic bottles and mounting them with rubber bands on long brass rods. Check them out to see where this project can go; I've posted a few videos at instructables.com/Upcycled-Bottle-Hang-Drum.

In the meantime, get ready to lose many hours playing this instrument, between improvising grooves and melodies, trying different tunings, and adding more bottles!

THAT CUTTING EDGE PLASTIC SOUND

YOUR 3D PRINTER CAN BE A POWERFUL MUSICAL TOOL

Written by Caleb Kraft

Paul Ellis (aka Polymate3D) has developed a high-quality speaker system you can 3D print from scratch. Ellis is determined not only to make his speakers from as many 3D-printed components as possible — cone, dome, spider, cabinet, etc. — but for them to actually sound fantastic. You just have to provide the magnets, wire, and other non-plastic parts necessary for a good speaker.

Ellis's previous versions are available for free at printables.com/@PaulEllisPolymate3D. Choose your own color scheme for the speaker and customize the enclosure to be exactly what you want. If you subscribe to his Patreon (patreon.com/polymate3d), you can get his latest and greatest.

Adobe Stock-Neo, Paul Ellis

Jim Axelsson, obert graham, mikolas zuza, jofoeus, cody, gringer

3D-PRINTED INSTRUMENTS

Feeling more of the musical spirit? It's time to start spitting hot plastic to make an instrument! There are so many to choose from, surely there's something that appeals to the bard inside you.

STRINGS

1 **THE JAX VIOLIN (DRAGON)** by JAx
printables.com/model/52215

2 **3D PRINTABLE HARP** by Grhmhome
printables.com/model/93081

WIND

3 **PLAYABLE OCARINA** by Mikolas Zuza
printables.com/model/65399

4 **PAN FLUTE** by jofoeus
printables.com/model/358559

PERCUSSION

5 **PERCUSSION FROG** by Wheelbarrow_of_Melons
printables.com/model/245546

MUSIC BOXES

6 **MUSIC BOX WITH HOLLOW CYLINDER**
by wizard23/gringer, printables.com/model/79444

- Customized: Harry Potter "Hedwig's Theme"
  thingiverse.com/thing:436043
- Customized: Beethoven's "Für Elise"
  thingiverse.com/thing:552498

SONIC PLAYGROUND

Written by Jennifer Blakeslee

JENNIFER BLAKESLEE runs the Global Maker Faire program and lives in Oakland, California. She has worked in events, radio, writing, education, big data, and public policy.

THESE CONTRUCTS PUSH THE LIMIT OF **MUSICAL INSTRUMENTS**

Exuberant, dissonant, melodic, mnemonic ... through sound we not only hear the world, we experience it differently. And although there is no shortage of instruments and genres with which to get your groove on, some music makers go the extra mile — not just playing instruments but creating them. The "sound engineers" below have made unusual and fascinating instruments that both bend the rules and, sometimes, reinvent them entirely.

1 KNURL CELLO

Rafaele Andrade

rafaele-andrade.com

Something special is bound to happen when the music, the performer, and the audience become a single feedback loop. Brazilian composer and cellist Rafaele Andrade is inventing ways to do just that. A pioneer in the field of applied experimental music, she has created Knurl, a reconfiguration of a baroque cello format — in which the performer can control up to six sound channels by changing sliders and pressing strings — into an interactive interface that shares control of the artistic experience between the audience and the performer.

2 NEEDLE NAILS

Victoria Shen

evicshen.com

Sometimes someone creates something so cool that it seems as if it has always existed. Cue San Francisco-based experimental sound artist Victoria Shen, aka Evicshen, and her Needle Nails. These artfully wired, any-color talons not only look the part — so glam they were lifted by Beyonce's creative team, garnering Shen a deserved public apology from the artist — they allow her to create the unique sounds + scratch she's become known for in her live performances.

3 NULL BEAM

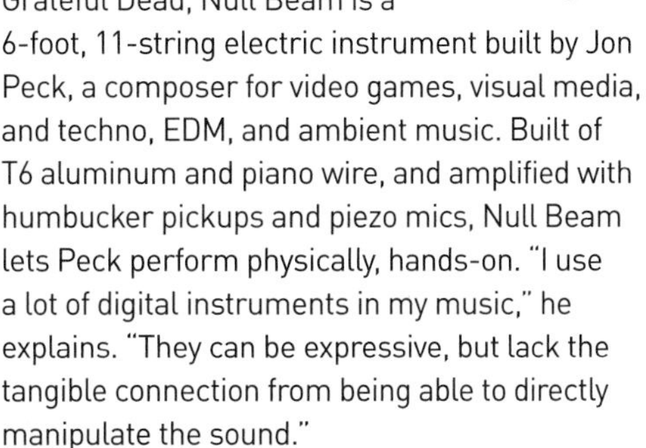

Jon Peck

null.band/beam

Inspired by Mickey Hart of the Grateful Dead, Null Beam is a 6-foot, 11-string electric instrument built by Jon Peck, a composer for video games, visual media, and techno, EDM, and ambient music. Built of T6 aluminum and piano wire, and amplified with humbucker pickups and piezo mics, Null Beam lets Peck perform physically, hands-on. "I use a lot of digital instruments in my music," he explains. "They can be expressive, but lack the tangible connection from being able to directly manipulate the sound."

4 COMMODORDION

Linus Åkesson

linusakesson.net/commodordion/index.php

Swedish maker Linus Åkesson really got creative with the fight against planned obsolescence. The Commodordion is an 8-bit accordion primarily made of vintage C64s, floppy disks, and gaffer tape. It's worth checking out his iterative process of 3+ years (which also produced the cleverly named Sixtyforgan and Qwertuoso), especially how he came up with a solution for the bellows: While watching a video on renewable energy, he heard wind from a turbine hitting the reporter's microphone and thought, "*That's* how I'm going to measure air flow!"

5 EGGIOPHONE

Hugh Jones

crewdson.net/eggiophone.html

Ever wondered what to do with those colorful, plastic eggs that kick around at Easter time? Musician and instrument builder Hugh Jones, aka Crewdson, painted them with black conductive paint, added a Teensy microcontroller, and ended up with a fully chromatic MIDI controller that can "play any sound in the world" when connected to software. Three potentiometer knobs and five momentary buttons are used to shift octaves and as gated loopers.

See and hear more unique and amazing makers at Maker Music Festival 2023, May 20–21 at makermusicfestival.com

Flip-Dot Animation

Written and photographed by Owen McAteer

How to control the awesome mechanical display that flips physical pixels, magnetically

Flip-dots are a 20th-century electromechanical display technology that uses a matrix of small magnetic discs which can be flipped between two states (usually black and white, or black and yellow) making up a dot-matrix display of pixels to write text, draw images, or even play animations and games.

Invented in the 1960s by Kenyon Taylor for airport signage, flip-dots (aka flip-discs) became popular in the 70s and 80s for stock market tickers and public information displays in airports and train stations, where they were used to show schedules, maps, and other information. Flip-dots were seen in highway signs and bus destination displays, and were often used for advertising, on billboards and in shopping centers, to display dynamic, attention-grabbing messages.

Today flip-dots have mostly all been replaced by LCD and LED displays, but there's something about the charm of a kinetic display that I love. And that satisfying, fluttering, clicky noise they make is just incredible. The physical movement and sound of each pixel flipping on and off is beautiful.

In recent years flip-dots have found rebirth in advertising, with hobbyists, and in art installations, most notably those of Breakfast kinetic art studio in New York. Today AlfaZeta (flipdots.com) in Poland is one of the few companies still manufacturing these displays with control boards. You can also find plenty of second-hand flip-dot displays on eBay but many lack control boards and/or documentation.

In my art work, I use AlfaZeta XY5 Flip-Dot boards in a variety of sizes. Their design and their control board allow for 30fps animation and can be controlled from a PC, Arduino, or Raspberry Pi. They take 24V for power and RS485 wiring for serial data using an 8-bit protocol.

TIME REQUIRED: 5–6 Hours

DIFFICULTY: Intermediate

COST: $400–$600

MATERIALS

- **XY5 Flip-Dot matrix display panel(s)** by AlfaZeta, flipdot.com. These are pricey, over $450 for a 28×14 panel, but you can find them used (see "Where to Buy?" below) or substitute similar panels.
- **Power supply, 24V, at least 1A per panel**
- **RS-485 converter** Cheap USB-to-RS485 converters are fine, such as Amazon B07CMY1DGK or AliExpress 3256802833469866. For faster framerates, use an ETH-RS485 such as UK-System ETH-UKW485SR140 (www.sklep.uk-system.pl) or Waveshare 15731 (waveshare.com).

TOOLS

- **Computer with software:**
  - **Processing** free from processing.org/download
  - **FlipDot Controller App** free from github.com/owenmcateer/FlipDots
- **Screwdriver**

OWEN McATEER is a creative coder and generative artist working under the moniker Motus Art, based in Madrid, Spain.

My free, open-source FlipDot Processing App (github.com/owenmcateer/FlipDots) makes drawing and animating on these displays as easy as possible, but some hardware setup is required; I'll show you how in this article. (The app can also run as a simulator, so you can learn to use it while you're waiting for your flip-dot panels to arrive!)

HOW DO FLIP-DOTS WORK?

Each disc contains a small permanent magnet and rotates on an axle. The disc aligns its face either up or down, by being attracted or repelled by the magnetic field of an electromagnet's stator pins produced by the coil beneath.

The polarity of the coil and electromagnet is flipped by a current pulse from a control board, which reverses the direction of the current through the coil, thus reversing its field and flipping the disc. You can see the whole set/reset cycle in the diagram on the following page.

How Flip-Dots Work

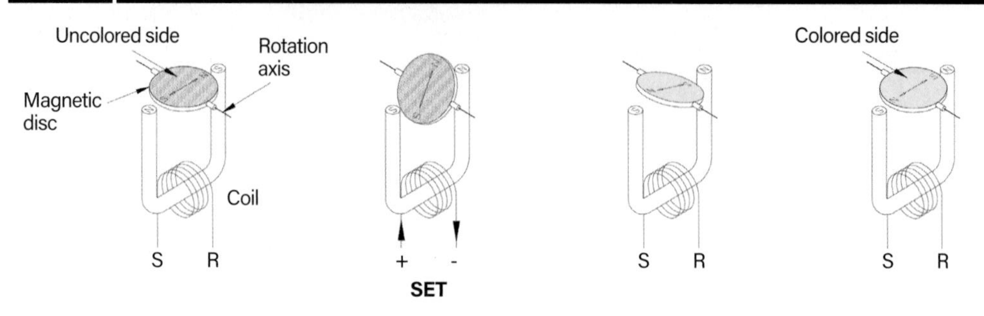

With a positive impulse on **RESET** the disc turns to the **uncolored** face.

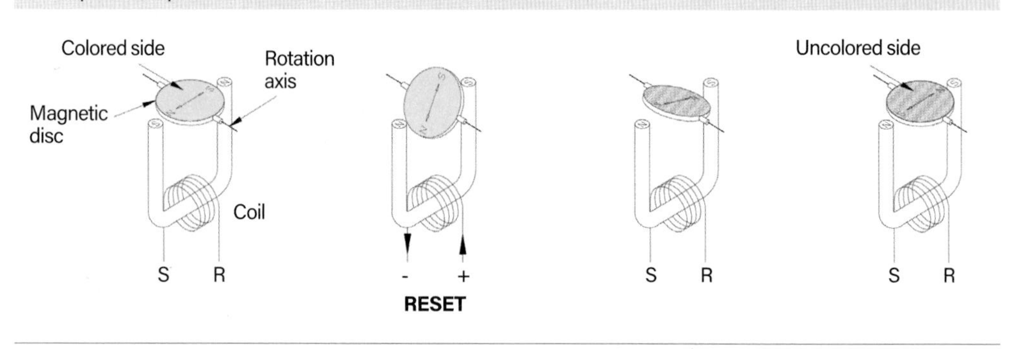

Adapted from Elettronica Digitale (eldisrl.com)

When power is disconnected the magnetic field remains in place to hold all discs in their positions, making flip-dots extremely energy efficient when not updating frequently. This is quite similar to early magnet storage drives, which were the inspiration for the flip-dot's inventor.

Each disc's coil polarization can be flipped in as little as 450μs (microseconds) making 30fps animation possible with the right control board.

WHERE TO BUY?

Flip-dot displays can be tricky to acquire but there are a few options.

AlfaZeta flipdots.com
These are the panels I use for my installations. AlfaZeta have created a modern flip-dot matrix and control board to enable high-framerate playback, all with a simple serial data protocol. My code for drawing and animating on flip-dot displays is all written with this protocol in mind.

Breakfast breakfaststudio.com
A Brooklyn-based artist studio that makes some of the most beautiful flip-dot art installations. They build custom panels but don't sell to the public; however they are clearing out a limited number of used AlfaZeta panels. Check on eBay at ebay.com/usr/breakfastny.

Flipo flipo.io
A Kickstarter campaign (bit.ly/Flip-disc-Kickstarter) set to launch in 2023 for Arduino-based flip-disc indicators and displays of various sizes, by a maker from Poland.

Decommissioned Flip-dots
eBay and other auction sites often list secondhand, decommissioned flip-dot displays taken from old buses and airport/road signage. This can be the cheapest way to obtain a flip-dot display but the condition and age, documentation, and whether or not it includes a control driver board could make it a challenging but fun build that requires some amount of hacking and reverse engineering.

DIY Components
A few companies still manufacture flip-dot

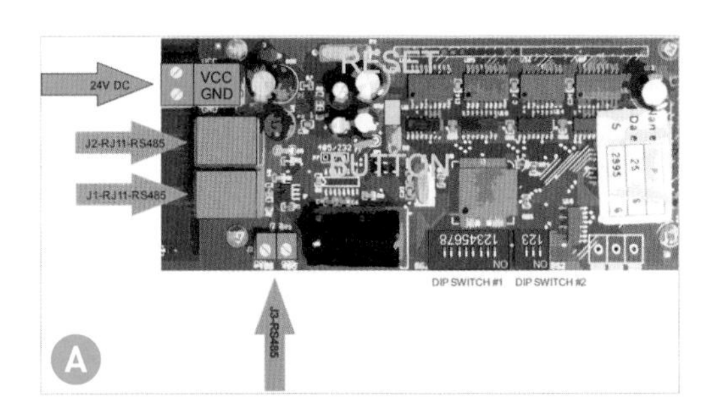

components. If you're up for a challenge you can source these components and build your own display and control board. Sadly all prices/quantities are on request only.

- **Elettronica Digitale (EL.DI.)** www.eldisrl.com/en-gb (Italy)
- **ScoreTronics** scoretronics.com (Illinois, USA)
- **Magsign** magsign.jp/english/products (Japan)

SET UP YOUR FLIP-DOT DISPLAY

This guide will cover all the hardware and software steps you'll need to get your own flip-dot display up and running.

1. WIRE IT UP

Connect your 24V power supply to the flip-dot control board's DC screw terminals.

Connect your RS485 converter from your computer to the control board with a JR11 plug or the screw terminals (Figure A).

2. SET THE DIP SWITCHES

Each controller has two DIP switches that need to be set: a 3-pin baud rate and an 8-pin panel address (Figure B).

Baud rate (3-pin DIP)

The communication transfer rate. I recommend the fastest value of 57600, as I found anything less

FLIPPIN' AWESOME

The greatest flip displays we've seen have got to be from Breakfast, an art studio in Brooklyn founded by three coders who wanted to explore how to manifest data kinetically.

Breakfast began with large, interactive electromechanical works commissioned by brands (like their beautiful *Thread Screen* for Forever 21), then moved into the fine art world with haunting flip-disc works that render climate change and deforestation data in real-time while interacting with viewers via motion tracking cameras.

Going flip-discs one better, they recently innovated their own large, tile-like "Brixels" that enable massive, stunning works like *Pulse*, reminiscent of sculptor Ned Kahn's shimmering kinetic facades except driven not by the wind, but by information.

—Keith Hammond

Breakfast, Owen McAteer

too slow to handle even 20fps. Set these switches to OFF-ON-ON. (The GitHub *readme* file has the configurations for other baud rates in case you need them.)

Address (8-pin DIP)
This is the address ID used when pushing out the image data. Stick with the default magnetizing time (500μs) unless you find a need to push it faster.

0–5 Address in binary code (natural)
6 Magnetizing time: OFF = 500μs (default), ON = 450μs
7 Test mode: ON/OFF. OFF = normal operation

3. SET UP SERIAL DATA

To send frame data from your computer to the display, you can use either Ethernet or USB. For small displays USB is fine but larger displays with good FPS will require an ETH solution. Download the project code from github.com/owenmcateer/FlipDots and open the file *FlipDot/config.pde*.

ETH converter — Connect the hardware: PC → ETH → ETH-RS485 converter → FlipDot panels. Set the following settings in *config.pde*:

- Set **`castOver = 1`**
- List all ETH convertor IP addresses and port numbers in **`netAdapters`**

USB converter — Connect the hardware: PC → USB-RS485 converter → FlipDot panels. Set the following settings in *config.pde*:

- Set **`castOver = 2`**
- List all USB converter COM port addresses and baud rate in **`serialAdapters`**

4. CONFIGURE THE FLIPDOT APP

Install Processing 4 (processing.org/download) for your system and launch *FlipDot/FlipDot.pde* (Figure C).

Now tweak *config.pde* for your setup:

- Make sure you've set your serial converter type as shown above.
- Set **`config_cast = true`** to cast data to the flip-dot panel. (Leave this as **`false`** to run simulator only.)
- Set the flip-dot panel and display settings:

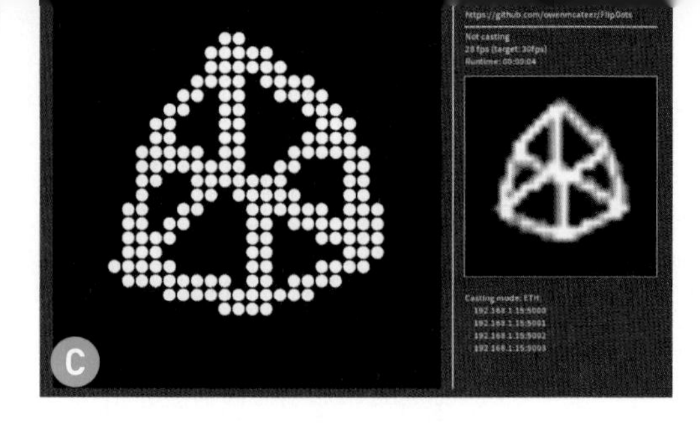

C

```
111  void createPanels() {
112    // Single 28x14 panel
113    panels[0] = new Panel(0, 1, 0, 0);
114    panels[1] = new Panel(0, 2, 0, 7);
115
116    /**
117     * Square display
118     * Made up of 4 stacked panels
119     *
120    panels[0] = new Panel(0, 1, 0, 0);
121    panels[1] = new Panel(0, 2, 0, 7);
122    panels[2] = new Panel(0, 3, 0, 14);
123    panels[3] = new Panel(0, 4, 0, 21);
124     */
125
126    /**
127     * Large sqaure
128     * 2x8 panels
129     *
130    panels[0]  = new Panel(0, 1, 0,  0);
```
D

E

`panels[0] = new Panel(0, 1, 0, 0);`
where the **`Panel`** parameters are:
1. Adapter ID (see **`netAdapters`** or **`serialAdapters`**)
2. Panel ID (set on the 3-pin DIP switch)
3. X-position in total display
4. Y-position in total display

For a single 28×14 panel, you can leave *config.pde* as it is. For more than one panel, add a **`panels[]`** config line for each panel you have.

- Finally, set the number of panels you have in the following line:
`Panel[] panels = new Panel[2];`

See *config.pde* for more examples and panel layouts (Figure D).

FLIP OUT!

Your flip-dot display is ready for shimmery, clickety action!

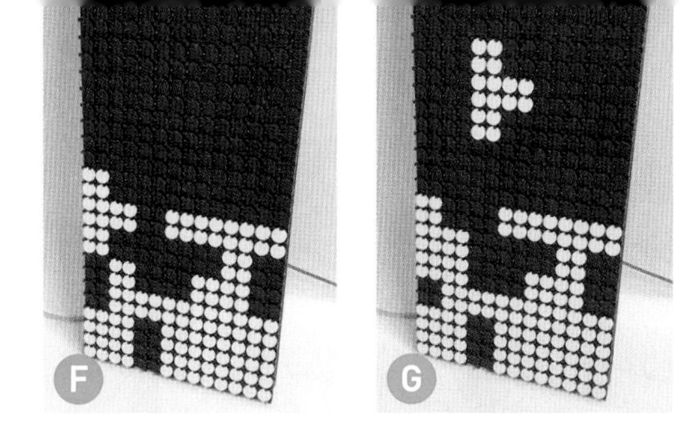

Hand control with Leap Motion hand tracker

Flip-dot display of Intel RealSense 3D depth camera

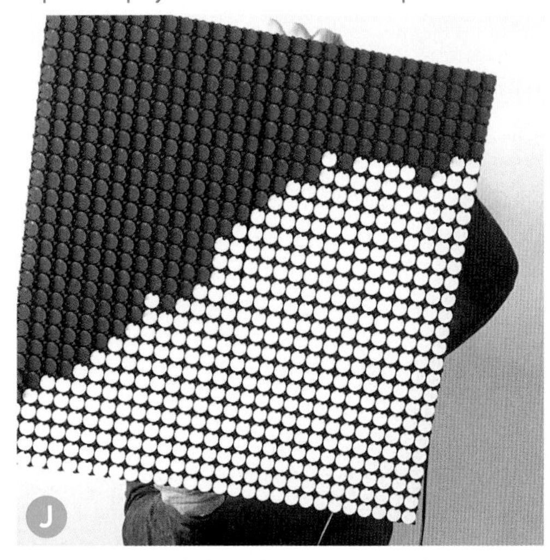

Accelerometer + Arduino for water simulation animation

FLIP 7-SEGMENT DIGITS

A similar product made by AlfaZeta are their XY7 Flip-Digit panels. These are your standard 7-segment digit design but with all the advantages and beauty of Flip-Dots. No LEDs here, just magnets and plastic. The control board uses the exact same protocol as the Flip-Dots XY5 so it is relativity easy to use; I've also published a control app for them (github.com/owenmcateer/FlipDigits).

These are perfect for displaying numbers (obviously), text (with some character limitations like K, M, X, Y), and images that take advantage of the linear geometry of the segments. You can also try to simulate grayscale with the segments in a digit-per-pixel setup, but you'd need a large display to pull this off.

Check out artist Ksawery Kirklewski (instagram.com/ksawerykomputery) for more amazing Flip-Digit art.

- **Coding animations** — Now you can create, draw, animate, whatever you want! Everything gets drawn to **`virtualDisplay`**. I recommend looking at *example_anim.pde* and *example_blips.pde* for some examples of coding animations and how to draw them to the virtual canvas (Figure E).
- **Games** — Add an input device or gamepad and make playable games. *Tetris*, *Snake*, and *Pong* all work great on low-res pixel displays like this. This implementation of *Tetris* (Figures F and G) can also be found on my Github page.
- **Interactive art** — For my art installations, I often make flip-dots interactive with touch-free devices such as the Leap Motion hand tracker (Figure H) and Intel RealSense depth camera (Figure I), or just a simple accelerometer (Figure J).

Whittled Wonder

Written and photographed by Sean Nolan

Sharpen your knife skills by carving a chain from a tree branch

Around October every year, the Seattle wind and rain bury our yard in huge drifts of leaves. Inevitably a few big branches come down, providing great raw material for a classic whittling project. Let's turn one of those branches into a wooden chain.

First, you're almost certainly going to cut yourself at some point. *Be really careful — hands contain some delicate parts!* Cut-resistant gloves can help, but most importantly keep your knives sharp. I use a fine stone to start, and a strop at least every half hour while I work. Sharp knives cut with less force, are less likely to slip, and cut cleanly when things go wrong (so, faster healing). They're also just way more satisfying to use.

Next, it's easier to carve green wood rather than dry; the moisture makes for less "tear-out" along the grain. That said, green wood will shrink and can be vulnerable to cracking as it dries. If you are working a piece over multiple days, storing it in a plastic bag between sessions will help hold the moisture content stable. Rubbing with mineral oil will protect a finished piece.

A branch about 2" in diameter and 6" in length makes three nice-sized links, a good number to start. Most folks start with a square milled blank — this is fine, but a straight fallen branch works great too. The only trick is to pay attention to the pith at the very center. Depending on the tree, this pith can be super-soft, more like packed brown sugar than wood. Since the center of the branch will make up structural parts of each link, I keep a bottle of thin CA or "Krazy" glue on hand. Whenever I expose a new bit of pith I soak it with the glue and let it dry before digging in, which fixes the material in place well.

CARVE YOUR WOODEN CHAIN

1. MARK THE ENDS

Draw a tic-tac-toe style cross on the end of the branch, with the middle square being about ½" or a bit more per side (Figure A). Use a ruler to extend the ends down the length of the branch, and then draw a matching cross on the other end. The two crosses should be oriented together, aligned as closely as possible.

2. CUT THE CROSS

Cut away the lengths along the corners, leaving you with a long X-shaped piece (Figure B). You can do this with your knife, but I recommend a rip-cut saw if you have one available.

TIME REQUIRED: 2–4 Hours

DIFFICULTY: Easy

COST: $0

MATERIALS:

- **A green section of branch, about 2" by 6"**
- **Cyanoacrylate (CA) glue** aka super glue

TOOLS:

- **Penknife or basic carving knife**
- **Knife sharpening tool: strop, whetstone, etc.**
- **Cut-resistant gloves (optional)**
- **Hand saw (optional)**

SEAN NOLAN is a longtime software guy who lives on Whidbey Island in Washington State and spends his days building stuff with code and driftwood. Stop by and say hello at his blog, shutdownhook.com.

A

B

3. CUT THE LINK OUTLINES

The first cuts are pretty easy since they're all on the outside — be sure to refer to Figures C and D as you work!

Measure and mark the length in four equal parts, and then:

3a At the halfway mark, cut a notch on one arm of the X and then in the same place on the opposite arm.

3b At each of the quarter marks, cut opposing notches on the other two arms of the X.

3c On the arms notched in Step 3b, remove the material from the notch to the end of the blank.

3d On the arms notched in Step 3a, make four half-notches at the ends of the arms. When you're done, you'll be able to see the outlines of the three links in your chain (Figure D).

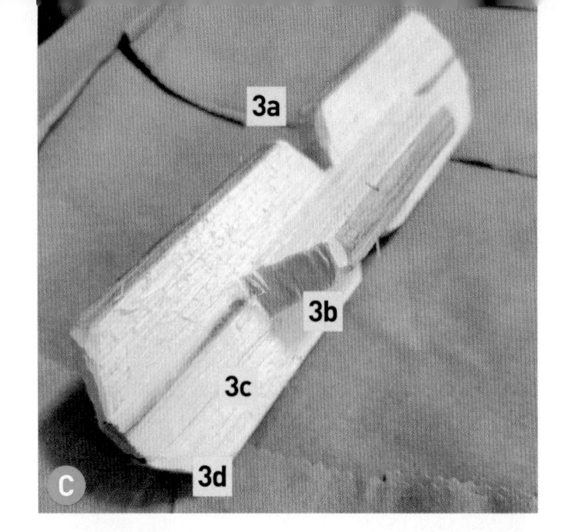

C

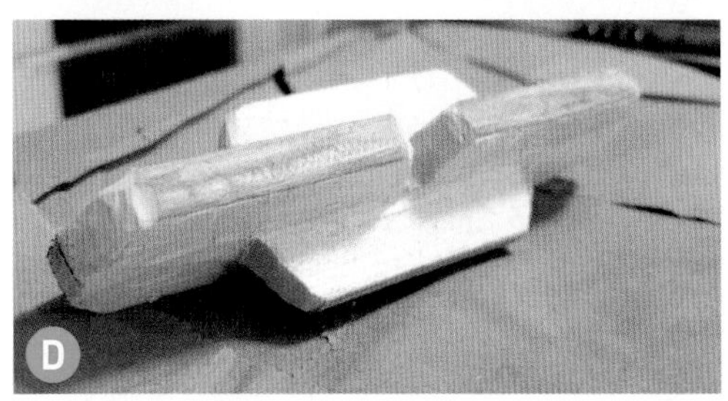
D

4. HOLLOW THE FAR ENDS

Hollow out the fully exposed centers of the two end links (Figure E). These internal cuts start to get more difficult. Make ***stop cuts*** along the outside of the target area and keep reinforcing them as you go deeper — they will prevent you from tearing out material beyond the center.

Take your time and don't try to remove too much material at once. I tend to use long ***V-cuts*** along the grain to pull out little toothpick-sized bits, alternating each side of the piece until I'm able to break through.

To learn more about woodcarving cuts, the best intro I've seen is the "4 Basic Cuts" video by Woodcrafter's Corner (youtu.be/82aZlw_Wfss). Other popular resources are carvingjunkies.com/wood-carving-cuts.html and woodiswood.com/all-7-cuts-to-master-in-wood-carving.

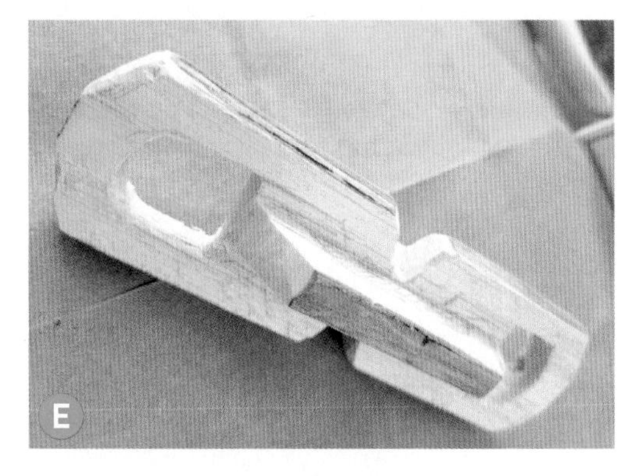
E

5. FINISH HOLLOWING THE END LINKS

Finish hollowing out the less-exposed centers of the two end links (Figure F). Use the same technique as in the previous step, being sure to keep strong stop cuts all along the outside. At this point the piece will start becoming pretty fragile, so be very aware of where you are applying force with your gripping hand and as you press in with the knife point.

When you've finished this step, the end links are clearly visible and almost complete.

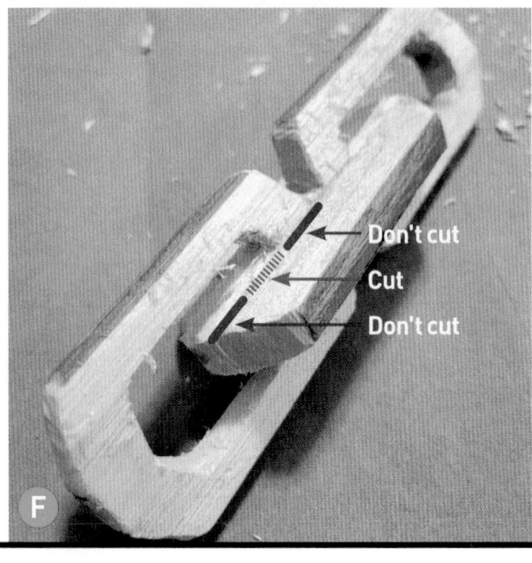

F

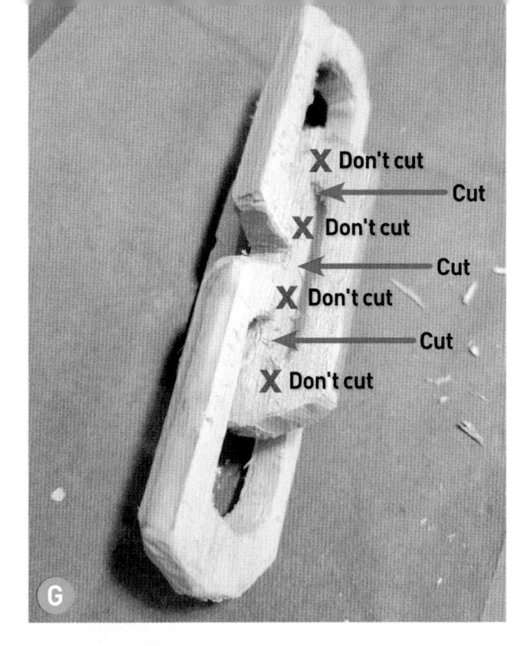

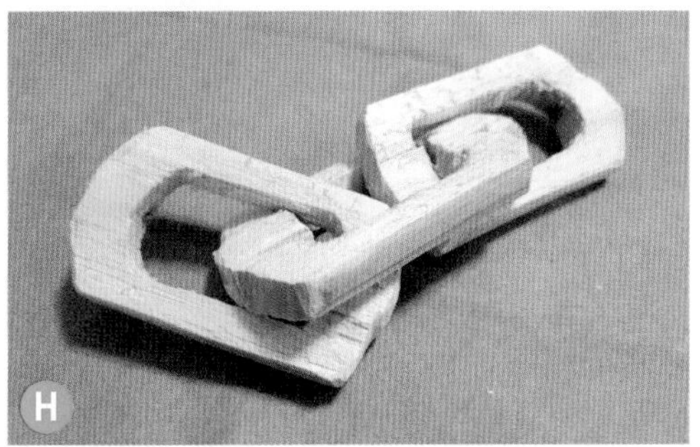

6. HOLLOW THE CENTER LINK

Hollow out the exposed long center parts of the inside link. These are shown about halfway done in Figure G. They are about the same narrow width as the hollows in Step 5, but they're longer and feel more awkward. Don't get impatient — take out small bits of material at a time, working on either side until the cuts meet. I find it easier to make deeper cuts towards the center of the piece until they meet, at which point there is more room to work the blade and clear the rest of the material.

7. FREE THE LINKS

This is the most exciting part! In this step you free the links from each other; in Figure G you'll see blue arrows pointing to the three areas to cut, and red X's for areas to avoid!

The piece will be very delicate, so make tiny "nibbling" cuts with the tip of the knife, being very careful to not apply too much force, especially with your gripping hand. Each of the three areas you need to clear will have four sides; work each side in turn until you break through and free the links (Figure H).

Resist the temptation to try to "snap" the remaining material when it gets thin; very little torque will break the link itself.

8. FINISH THE LINKS

Almost there! Where you've separated the links, there will be extra material to remove before profiling the links into a more regular, rounded form (Figure I). This is far less awkward and risky than Step 7, because now you can move the links around to create access to any part you need to work on.

Sand the links if you like, and apply a bit of mineral oil to protect the wood.

CHAIN REACTION

We made it — and the finished product is pretty awesome! Everybody seems to love a hand-carved chain, whether as home décor or just because they're cool. Nothing super-complex, but great whittling practice that requires patience and focus, both of which I can always use more of. And I do love working with materials that come from where we live. Fun stuff.

Retro-Futuristic LED Ring

Written and photographed by Charlyn Gonda

CHARLYN GONDA is a coder by day, maker by night from Alameda, California. She loves to create delightful, often glowy things that bring a bit of joy into the world.

TIME REQUIRED: 2–3 Hours

DIFFICULTY: Easy/Moderate

COST: $15–$20

MATERIALS

- **LED filaments, 26mm long (3)** Adafruit 5505. These are the non-flexible kind that you see in retro Edison-style bulbs.
- **Coin cell battery holder, CR1220, SMD type** Digi-Key 36-1072-ND
- **Brass rod, 3/64" (1.19mm) diameter, 12" long** K&S Metals 8161
- **Coin cell battery, 1220 size**

TOOLS

- **Soldering iron and lead-free solder**
- **Solder fume extractor**
- **Water-soluble flux** such as Chemtronics
- **Small detail paintbrush**
- **Third hand tool**
- **Big Wrapper Pliers, 13/16/20mm** Beadsmith XTL-5039
- **Needlenose pliers**
- **Flush cutters**
- **Fine sandpaper**
- **Steel wool**
- **Optional: Kapton tape, ring mandrel with ring sizes, ring sizing kit, alligator clips**

Make a mysterious, sci-fi inspired, glowy ring using LED filaments

I've noticed that some of the newer sci-fi films and TV shows are less about metallic shiny silver spaceships, and more about organic earth tones and devices with hints of gold. Instead of blocky LED bars, modern sci-fi props have a mysterious glow, with designs that look like they could be found in an archaeological dig.

This ring was a product of some experimentation and was somewhat of an accident, but I absolutely love how it feels like something that came out of the 2021 *Dune* movie. It's a simple build but requires some careful soldering — because the brass ring itself conducts the low voltage that lights up the LEDs.

1. BEND TWO V-SHAPED RINGS

1a. Cut one brass rod in half — this is probably too much length but you'll want to be able to customize it to your finger size (Figure A).

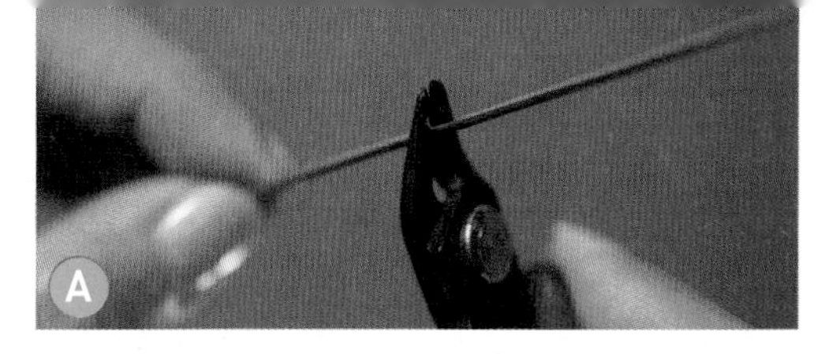

1b. Using the needlenose pliers, bend the rod around the halfway mark to about 90°. This will be the V shape on the front of the ring (Figure B).

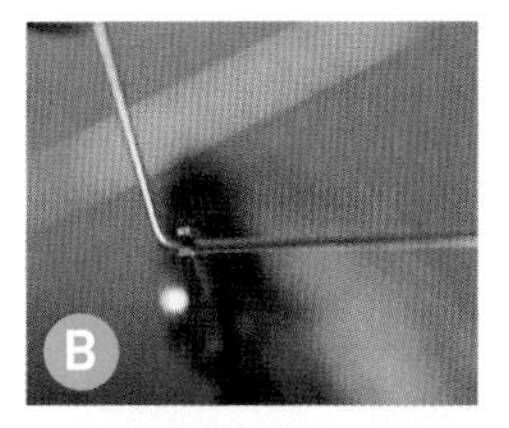

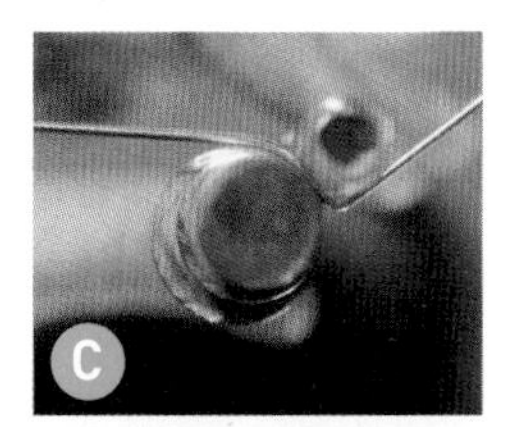

1c. Hold the bent rod in between the Big Wrapper Pliers, aligned with the 20mm barrel. Bend a curve outward on one side, so that it forms half of the V shape (Figure C).

1d. Do the same on the other side to complete the curved V shape. At this point, you can form the brass by hand so that the straight lines on either side of the V will form a straight line with one another (Figure D).

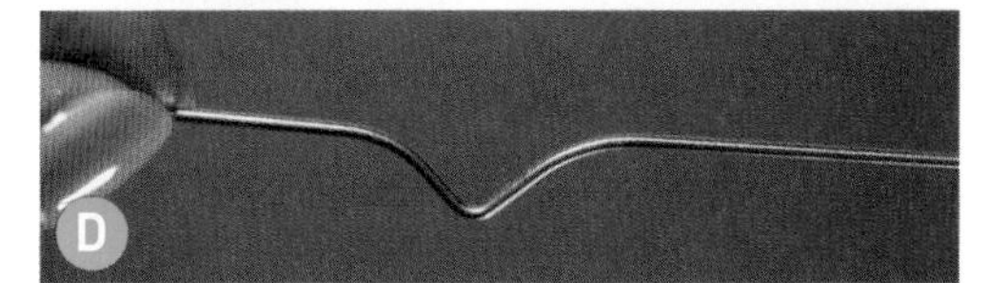

1e. Place the V in between the 20mm barrel and the other side of the wrapper pliers. Bend both sides (Figure E). The wire will form a diameter slightly larger than 20mm, but this will give you a good shape to start customizing to your ring size.

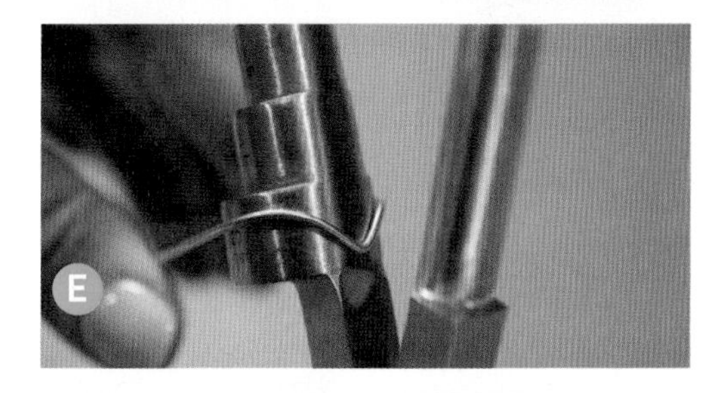

1f. Using either a ring mandrel or your own finger, form the circle so that it fits your ring size.

If you're using a ring mandrel, you could use a ring sizer kit to figure out what size is most comfortable and then use the ring mandrel to get to the desired size. I recommend sizing up at least a half size so that you can account for the thickness of the LED filament (Figures F and G).

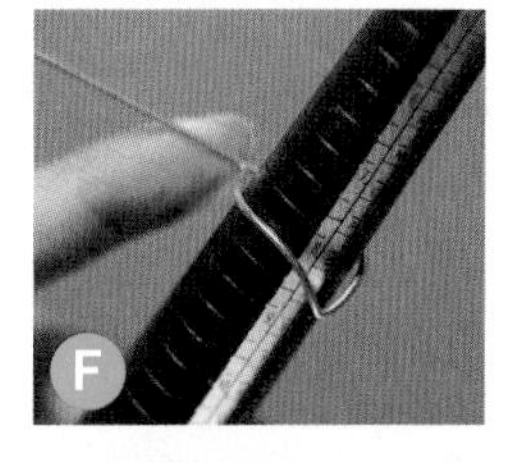

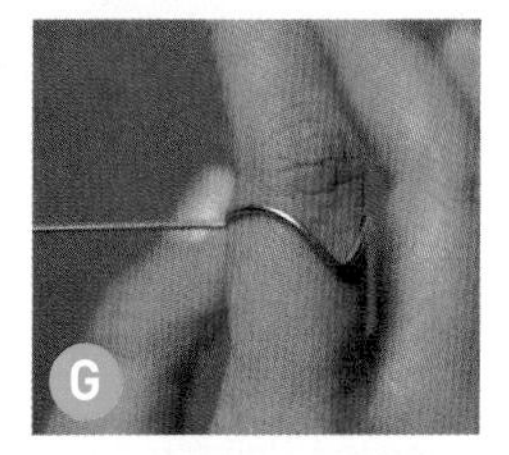

1g. Cut the excess brass rod, taking care to use the flat side of the cutter to achieve a flat cut (Figure H).

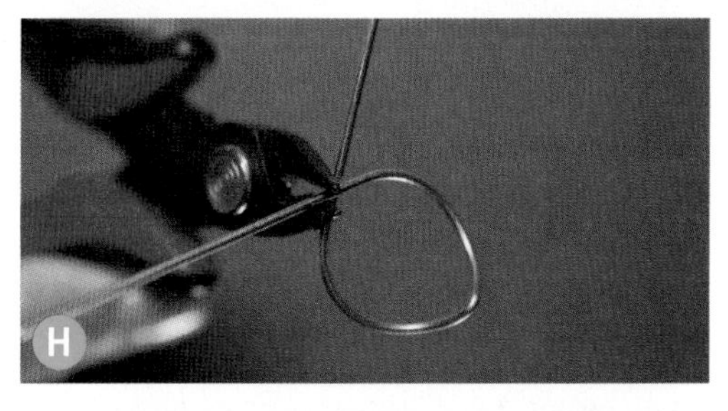

1h. Professional jewelers will think this is crazy, but I'm going to say it: It's OK if you have a gap in the ring! Take as much or as little time as you want to close this gap (Figure I).

Now do steps 1b–1g again to get two of the same V-shaped ring.

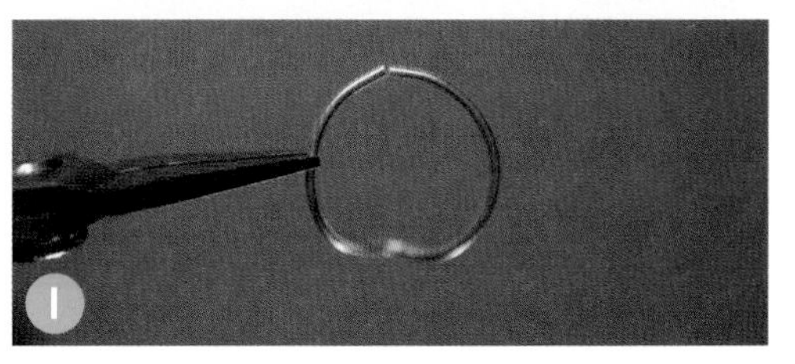

2. SOLDER TOP RING TO BATTERY HOLDER

2a. Using your third hand tool, position one V-shaped ring under the solder pad that touches the negative (–) side of the battery, roughly perpendicular to the battery holder (Figure J). This can be tricky, but take your time — remember that this is an art, not a science.

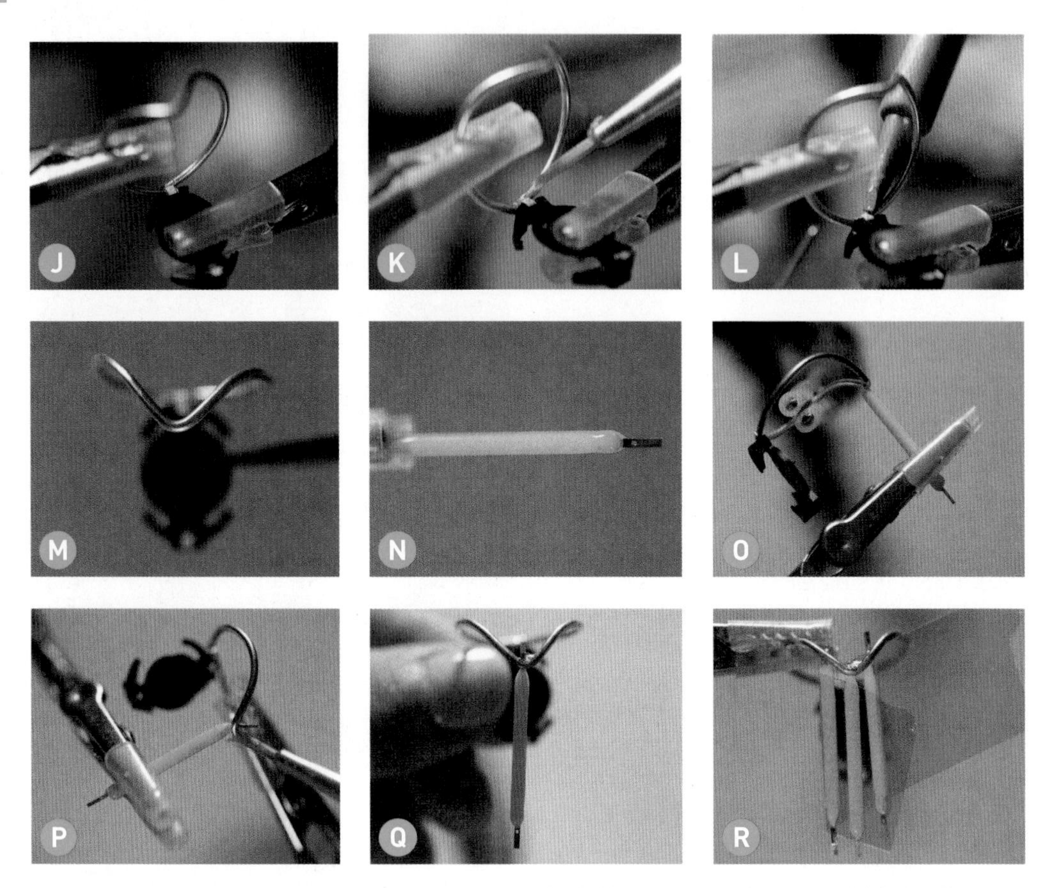

2b. Brush on a tiny bit of solder flux in the joint between the rod and the battery holder's solder pad (Figure K). This will help the solder adhere to the brass rod better. Using water-based flux will help with easier cleanup since you can just rinse the ring after soldering.

2c. Heat up the joints, and wait for a bit of the flux to heat up and release smoke. When you think it's hot enough, touch the solder to the joints themselves and not the soldering iron (as much as possible) — this will help to ensure a strong connection. Let the solder flow, and release (Figure L).

2d. Ideally you'll end up with the V shape roughly centered on the battery holder (Figure M). Also, it's OK if you don't end up soldering the gap shut, the alignment is a bit more important.

3. SOLDER THE LED FILAMENTS

3a. Identify the polarity of the LED filament. Each end has a solder pad, and the solder pad with a hole is usually the positive (+) side. Currently, the top piece is soldered to the negative (–) side of the battery holder, so you want to solder the ring to the pad without a hole (Figure N).

> **WARNING:** These LED filaments are quite fragile and will break if not handled carefully. They might still light up, but will flicker if broken.

3b. Slightly bend the LED filament's solder pad and position it against the center of the V, making sure the solder pad is on the inside of the ring, and position the filament so that it's roughly parallel with the battery holder (Figure O). Take your time here, it can be a little challenging to sufficiently align the filament. Try to get the brass rod as close as possible to the LED filament so that much of the solder pad is hidden behind the brass rod.

3c. Dab a bit of solder flux in this joint (Figure P), then heat both the rod and the solder pad with the soldering iron, and patiently wait for the solder to melt.

3d. Double check that the filament is aligned well enough (Figure Q). Remember to be kind to yourself — you're making a unique object!

3e. Do steps 3a–3c for the other two filaments, positioned on either side of the center filament. You might consider using heat-resistant Kapton tape to help you align both sides (Figure R). Keep going, you're doing great!

3f. At this point, you can test that your solder joints are good and the orientation is correct, using alligator clips and a 3V power source. Here I'm just using two alligator clips connected to a 2032 coin cell battery to make sure each filament will light up properly (Figure S).

4. SOLDER THE BOTTOM RING

4a. Using the same incredible maker tenacity you've been demonstrating this entire time, carefully position the bottom ring so that the front V touches the ends of the LED filaments and the backside touches the battery holder's solder pad (Figure T). This can be tricky; again, take your time and try not to break any of the filaments.

4b. It's OK if there's a small gap between the ring and the battery holder pad (Figure U) — you should be able to fill this gap with solder. Dab some flux on both the battery holder's solder pad and the brass rod (Figure V). Heat up both the rod and the pad, and apply solder.

4c. Bend each filament's solder pad outward and ensure that they stay inside the ring (behind the brass rods). Three more solder points with more flux here, and you're done (Figure W)! You can use some fine-grit sandpaper to smooth any pointy bits, and steel wool to get the brass to shine.

Now you can put the battery in the holder and see your creation glow!

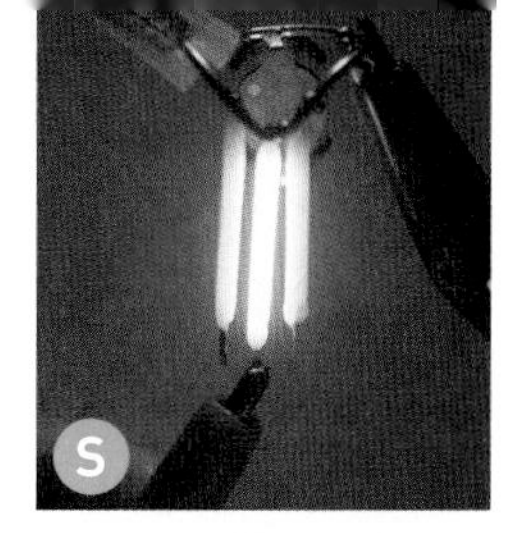

S

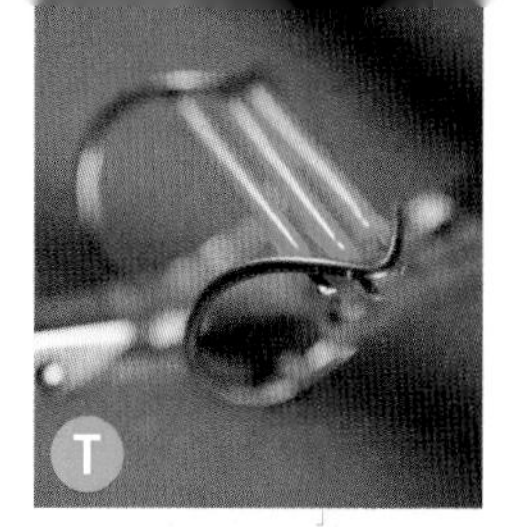

T

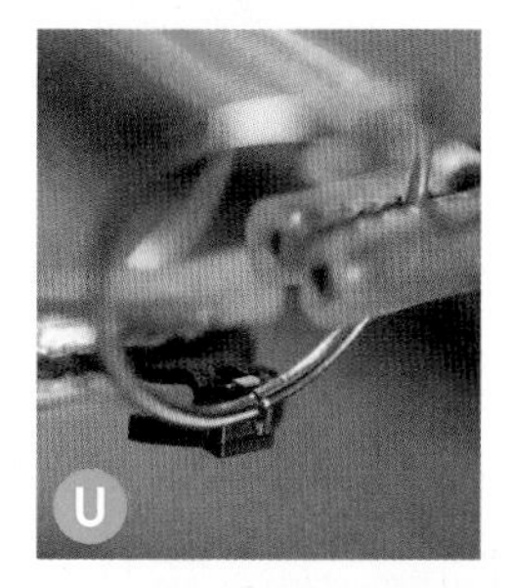

U

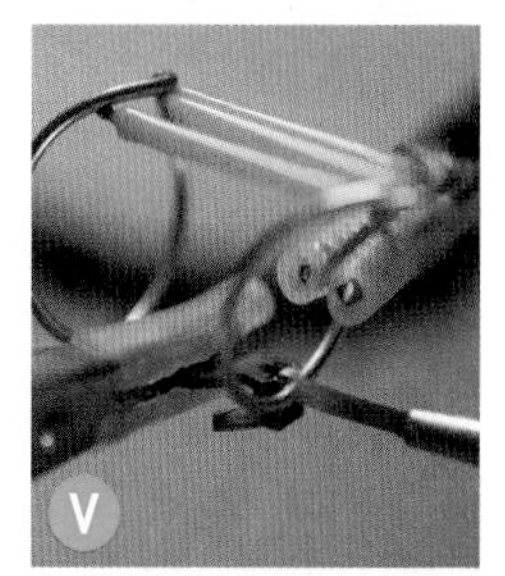

V

W

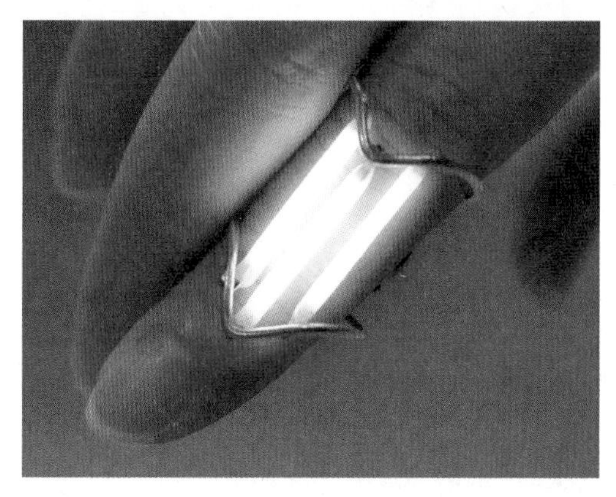

SCI-FI SPLENDOR

This ring is elegant enough for a glam night out, but I've also worn it casually around my workplace. A single 1220 coin cell battery should last the entire day — my last one went 15 hours, so I had to pop out the battery because I didn't want to waste the energy when I went to bed.

I love being able to incorporate a bit of glowy magic into my everyday, and I hope this guide helps you to make a little retro-futuristic, sci-fi inspired magic of your own!

Weird Resistors Part 1

Play With Your Food

Build an Atari Punk synth circuit — then get fruity with it!

Written and photographed by Lee Wilkins

LEE WILKINS is an artist, cyborg, technologist, and author of our "Squishy Tech" column in *Make:* looking at technology and the body and how they intertwine. Follow them on Twitter @leeborg_

We're going to talk about weird resistors. Yes, you've heard me. The resistor is one of the most underrated components in electronics. This article is the first of a series about weird resistors. True to our theme of Squishy Tech, in this issue we'll start by playing with our food.

WHAT IS A RESISTOR?

Electricity is measured by three quantities or properties: voltage, current, and resistance. If you've spent any time working with electronics, you've heard of ***Ohm's law,*** which says that the electric current through a conductor is directly proportional to the voltage, and the resistance is used to control that relationship. This can be described mathematically as:

- **Voltage = current × resistance**
- **Current = voltage / resistance**
- **Resistance = voltage / current**

This means that by knowing the value of two components, you can determine the third.

Controlling resistance is key, and the resistor is the component that does it. It's something you will need to do in so many circuits to limit the current and prevent your components from burning out.

Often resistors are made by winding a resistive wire around an insulated form and covering it with an insulator, or by applying a thin layer of resistive material onto an insulated form. These techniques ensure that the resistor is — roughly — the right value.

We can check the value of a resistor a few ways. The most common is to read the colored lines painted on most resistors — check the "resistor color code" online or in your Maker's Notebook (makershed.com/collections/notebooks-writing).

Another way is to use a multimeter to read the value. Just turn the knob to the resistance range you desire to read on the ohms (Ω) portion of the dial and touch the leads or probes to either side of the resistor. You'll notice that it might not measure exactly the indicated value; that's normal, as most resistors have a tolerance of 5% or 10%, and some can vary up to 20% (Figure A).

Resistor values are organized into a standardized series of values, the ***E-series,*** spaced so that their tolerance ranges do not overlap. So not all resistors are created equal!

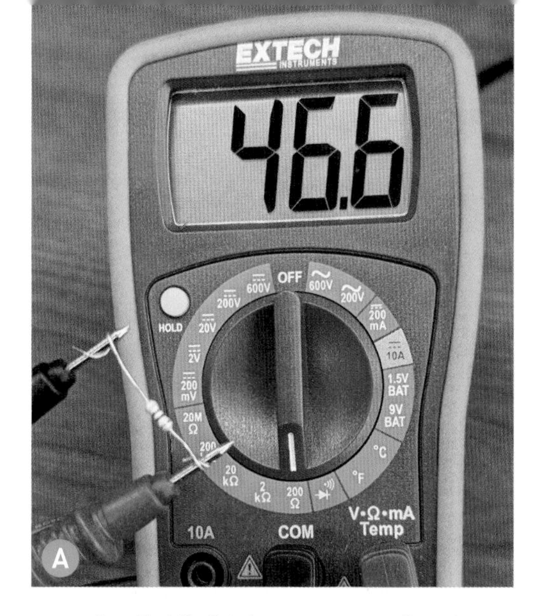

You can buy a package of resistors with a variety of values and use them for all kinds of projects. Using resistors to control the properties of a circuit can change the output, or be used to make decisions.

WHAT'S A VARIABLE RESISTOR?

It's easiest to think of resistors as having a static value. But we actively use ***variable resistors*** all the time in the form of knobs or slides. As you twist or move the knob, the resistance changes. If you've ever turned up the volume using a knob, you're a variable resistor user!

These variable resistors, known as ***potentiometers,*** work mechanically, by moving a conductive piece along a resistive surface, forcing the electricity to go through more or less of the resistive material. We'll explore this mechanic more in Weird Resistors Part 2.

Any material that conducts electricity also has a resistance; even copper wire has a very low resistance that can accumulate over distances.

EXPERIMENTING WITH RESISTANCE

Now it's time for the fun! I've experimented with a variety of fruits and foods to better understand their resistance, and how we can play with them inside circuits. You'll notice that dry foods will have a very high resistance (water is a pretty good conductor), and depending on the qualities and freshness of the food, the values can vary

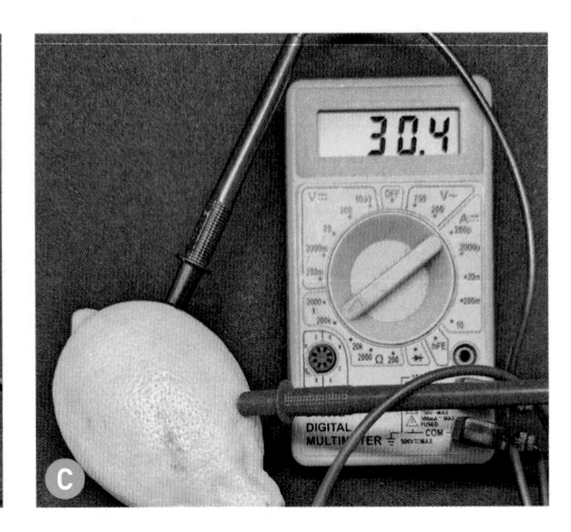

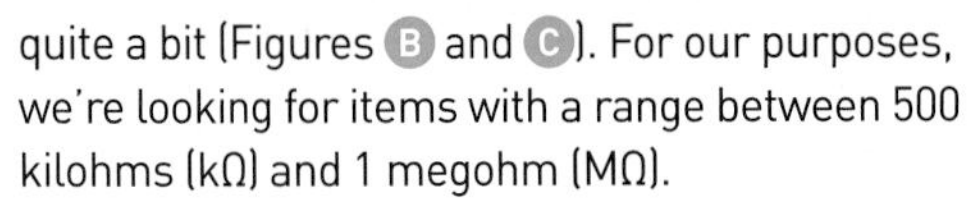

quite a bit (Figures B and C). For our purposes, we're looking for items with a range between 500 kilohms (kΩ) and 1 megohm (MΩ).

What's fun about using food or plant matter, is that as time progresses you'll notice the values change. If you leave your circuit connected, or try it again every few days, you'll see the values usually decrease as the living matter decays.

A 555 AND ATARI PUNK

The 555 timer is an infamous chip, used to make so many different circuits. One of my favorites is the Atari Punk circuit, which is basically a really simple synthesizer made from an oscillator. It's made from either two individual 555s, or a 556, which is just two 555s jammed into one neat package. The first Atari Punk circuit was designed in 1980 by Forrest Mims who called it a Sound Synthesizer and later the Stepped Tone Generator (Figure D). It was eventually renamed the Atari Punk Console by Kaustic Machines (Figure E), but this circuit has captured the heart of artful circuit makers everywhere.

How does this clever circuit produce sounds? We have to understand a few things about integrated circuits (ICs). ICs have numbered pins that perform different functions that allow us to configure the chip, for example this circuit configures the 555 chip in ***monostable multivibrator*** mode. When we trigger pin 2, pin 7 begins to charge a capacitor. This charging process sets off a high pulse on pin 3, causing the capacitor to recharge through an external resistor. The resistor limits the speed that the

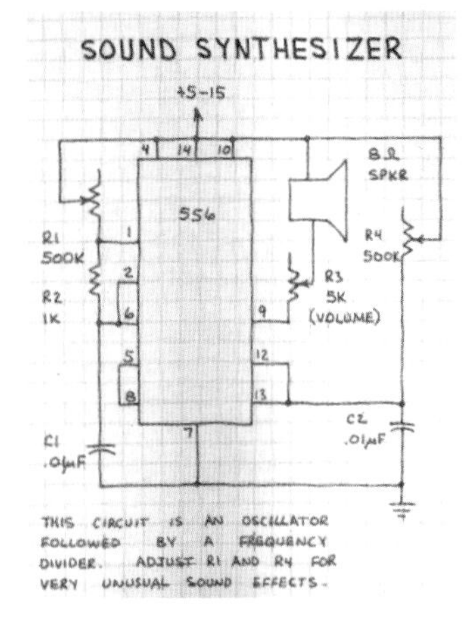

D

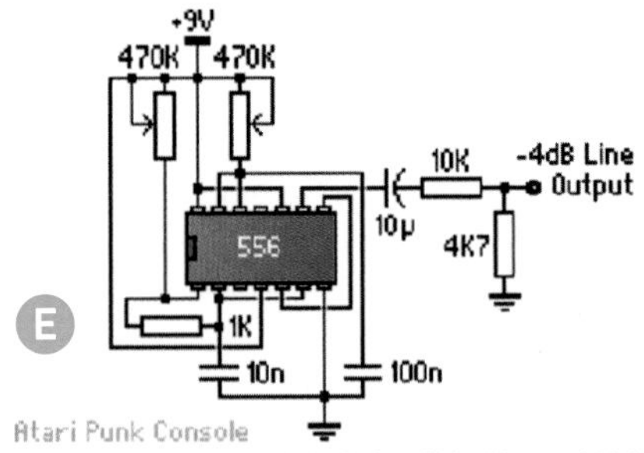

E

Atari Punk Console
kaustic machines - original circuit by Forrest M. Mims, III

capacitor can recharge, and thus determines the length of the pulse. The repeating pulse creates an audible sound through the capacitor connected to the speaker. By combining two 555 circuits we can create a more complex sound.

In our circuit, the resistor that limits the charge speed can be replaced by anything we want that has a resistance within the range we're looking for. So why not fruit? Let's get started!

Forrest Mims, Kaustic Machines

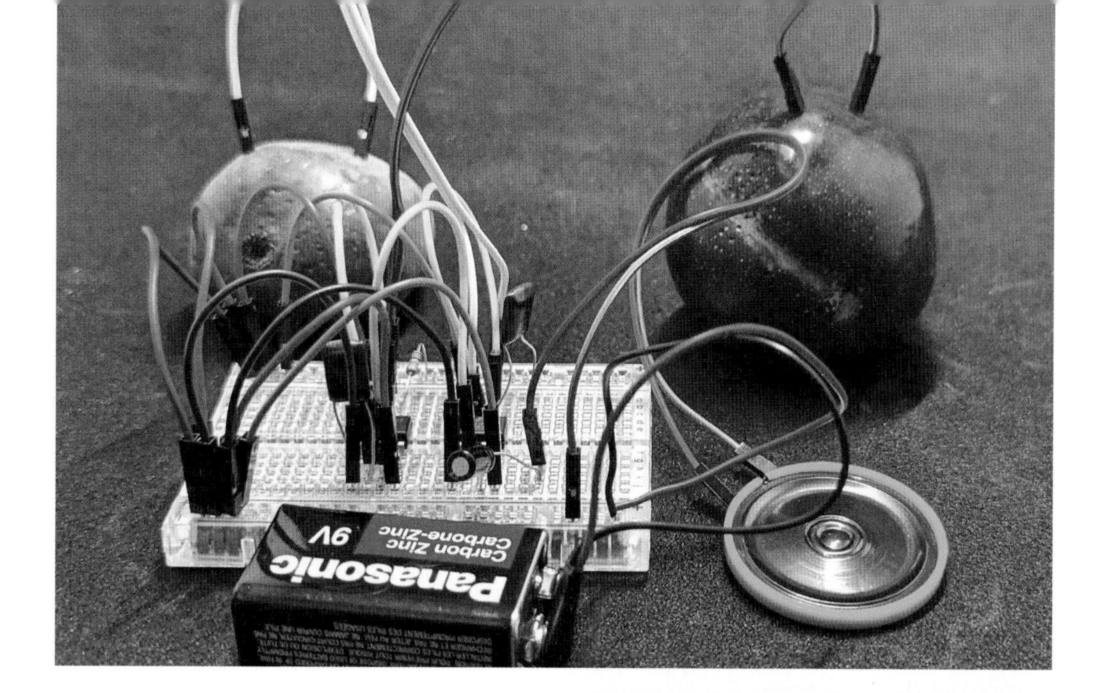

Build your Fruit-Controlled Atari Punk Circuit

TIME REQUIRED: 1–2 Hours

DIFFICULTY: Easy

COST: $20–$30

MATERIALS

- » **Solderless breadboard**
- » **Jumper wires, male-male**
- » **555 timer IC chips (3)**
- » **Resistors: 1kΩ (1) and 4.7kΩ (2)**
- » **Capacitors: 0.1µF (1), 0.01µF (1), and 10µF (1)**
- » **Potentiometers, 500kΩ (2)**
- » **Speaker, 8Ω or 4Ω**
- » **9V battery snap** with wire leads
- » **9V battery**
- » **Fruits of your choice (2)**

TOOLS

- » **Multimeter (optional)**

1. SET UP YOUR BREADBOARD

We're going to build the circuit in two parts. First, let's build the base by setting up your breadboard Place your two 555 chips in the center of the board, so that they straddle the center channel (Figure F). This ensures that the chip's pins — its metal legs — aren't connected to each other.

It's also important to keep in mind the orientation of the chips so that you know you're connecting the right pins. Check the top of the chip for a dot in the corner or a half-circle notch on the end. Make sure the two chips are facing the same way. As oriented here, with the dot or notch on the left, pin 1 is the bottom left pin, with numbers going across the bottom, left to right, and then reverse across the top, as indicated in the schematic diagram, Figure G on the following page. Confusing, but at least they're all the same!

If your breadboard has two power rails, marked with + and – symbols, at both the top and bottom, it's good to connect these using blue and red

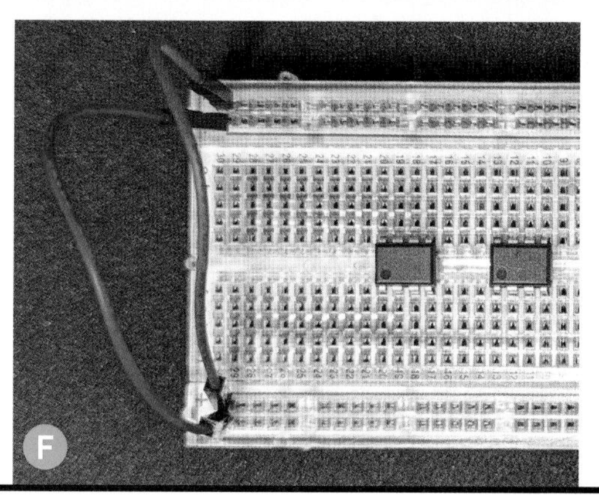
F

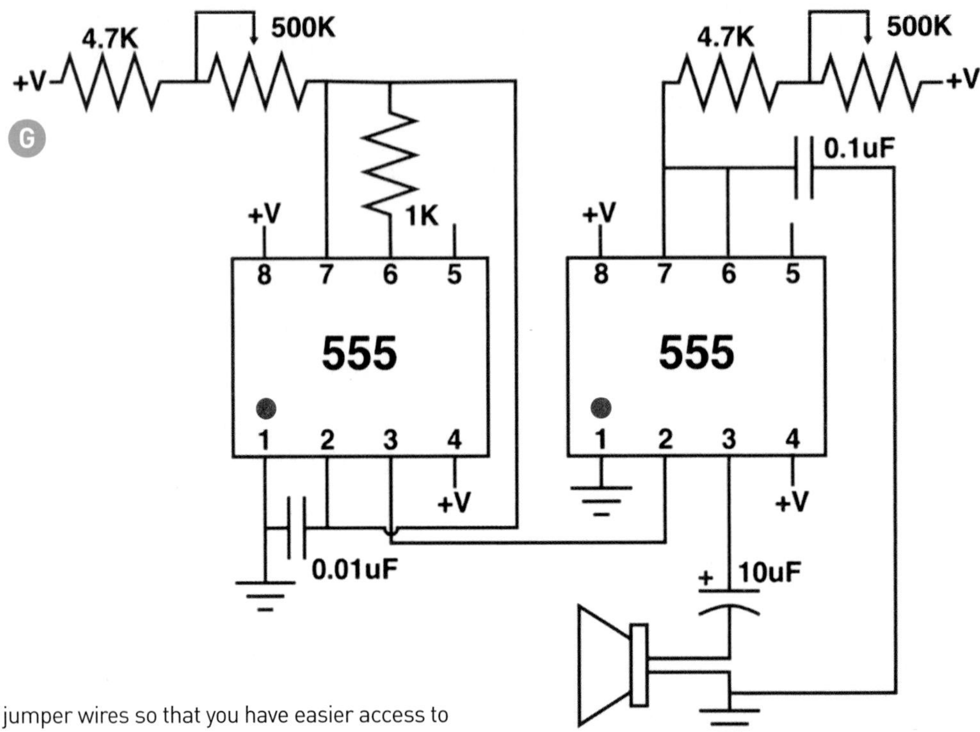

jumper wires so that you have easier access to power and ground. Then you can plug in your 9V battery snap, red wire to a positive rail and blue to a negative rail.

2. BUILD YOUR CIRCUIT

Now you'll follow the schematic diagram to assemble your circuit. Ultimately it should look something like Figure L, but yours might be arranged slightly differently. It's important to know that the diagram is not a representation of how the circuit looks, but how it is connected. The size of your breadboard and your comfort with electronics can be factors in how you assemble a circuit. I like to print the diagram out and cross out the lines as I connect them.

I'll usually start with the power and ground pins, 8, 4, and 1 (Figure H). Then connect pin 3 from the first chip to pin 2 on the second (Figure J). Here I also connected the first chip's pin 2 to its pin 7, and the second chip's pin 6 to its pin 7 (Figure K).

Then, I'll connect all the capacitors (Figure L), then all of the resistors, then anything else like the speaker. It doesn't really matter what order, some people connect the pins in numerical order, but having any kind of system can help you make

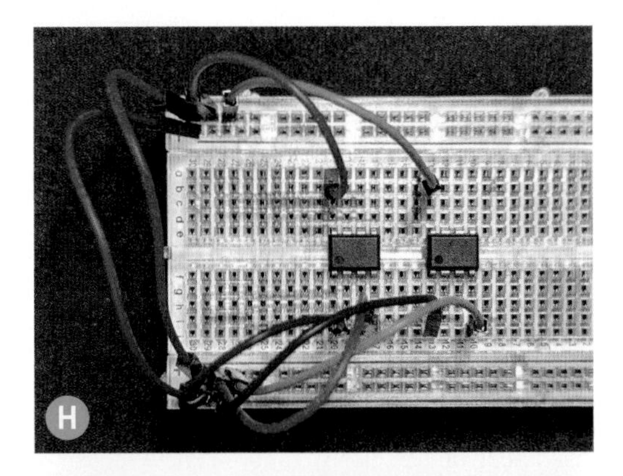

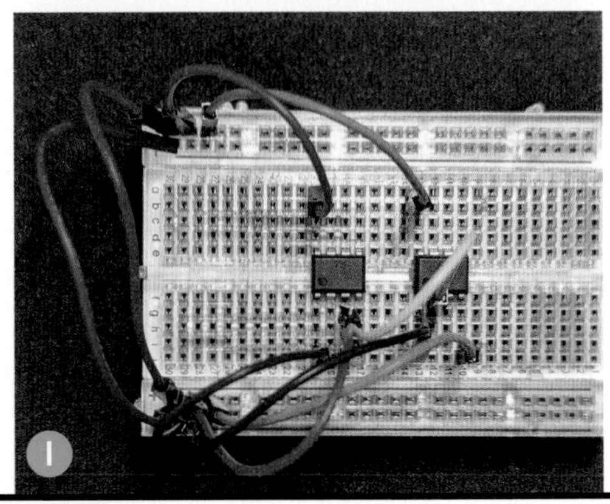

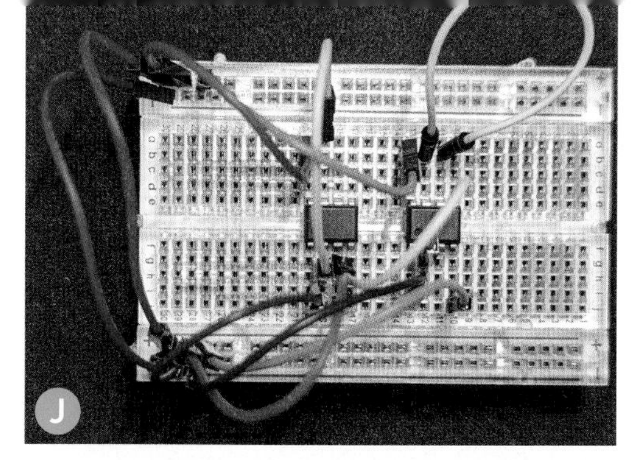

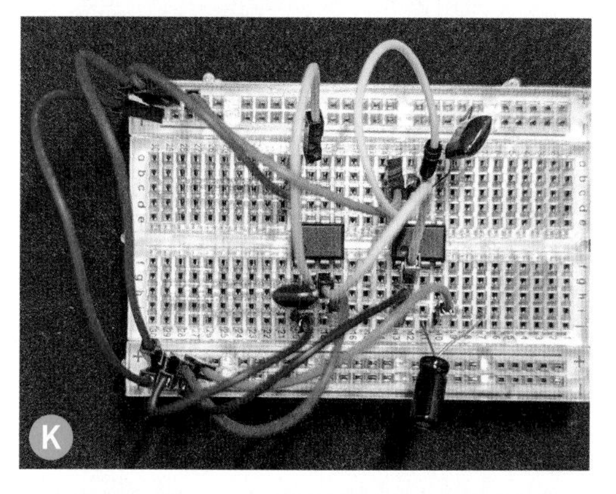

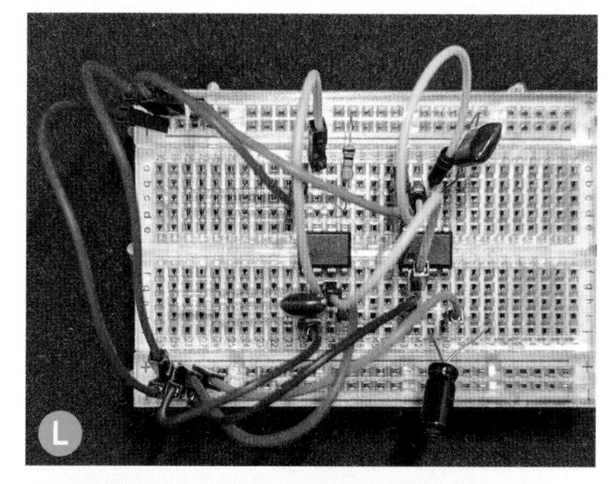

sure everything gets connected.

If you don't know how to read circuit diagrams, it's not as hard as it might look. The pins of the chip are numbered, and lines indicate what is connected to them. Any hole in a numbered row on the breadboard will connect with the pin that's in that row. V+ indicates a connection the positive (+) power rail, and the arrow-shaped symbol indicates a connection to ground, or the negative (–) rail. The zig-zag symbols indicate resistors; just make sure to add the corresponding resistor value to the right pin. The resistors with tiny arrows pointing at them are the potentiometers, and the symbols that look like two lines are the capacitors.

3. TEST YOUR CIRCUIT

Before adding any fruit, you should be sure your circuit works! Connect your 9V battery to the battery cap and twist the knobs. You should hear some great glitchy sounds. If your circuit isn't working, go through the diagram pin by pin and make sure that nothing is touching anything it shouldn't, and that everything is plugged in tightly to the breadboard. You can use your multimeter to test the ranges you enjoy most, and seek out some fruits that are in those ranges.

4. APPLY FRUITS!

When you're ready, remove the potentiometers and 4.7K resistors and replace them with two jumper wires, as shown in Figure M.

You can now plug them into any fruits you desire and listen to the sounds they make!

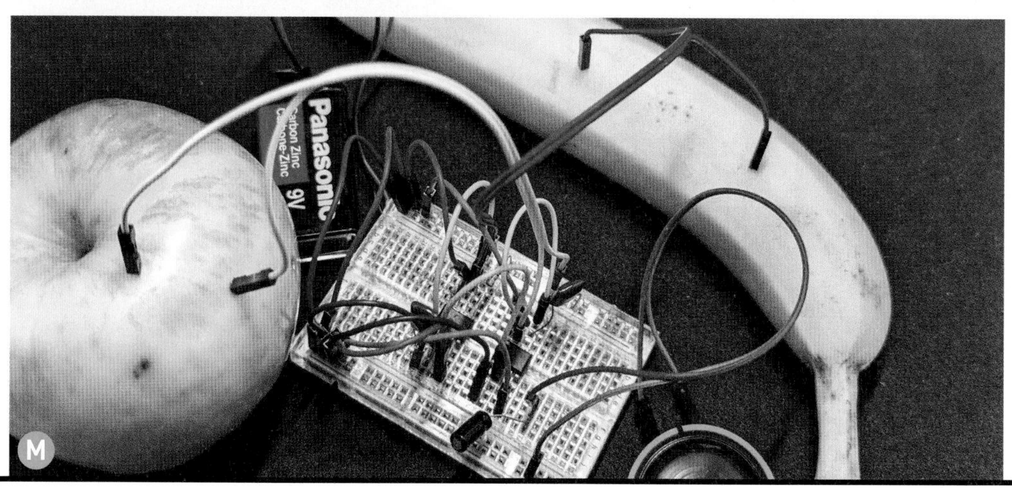

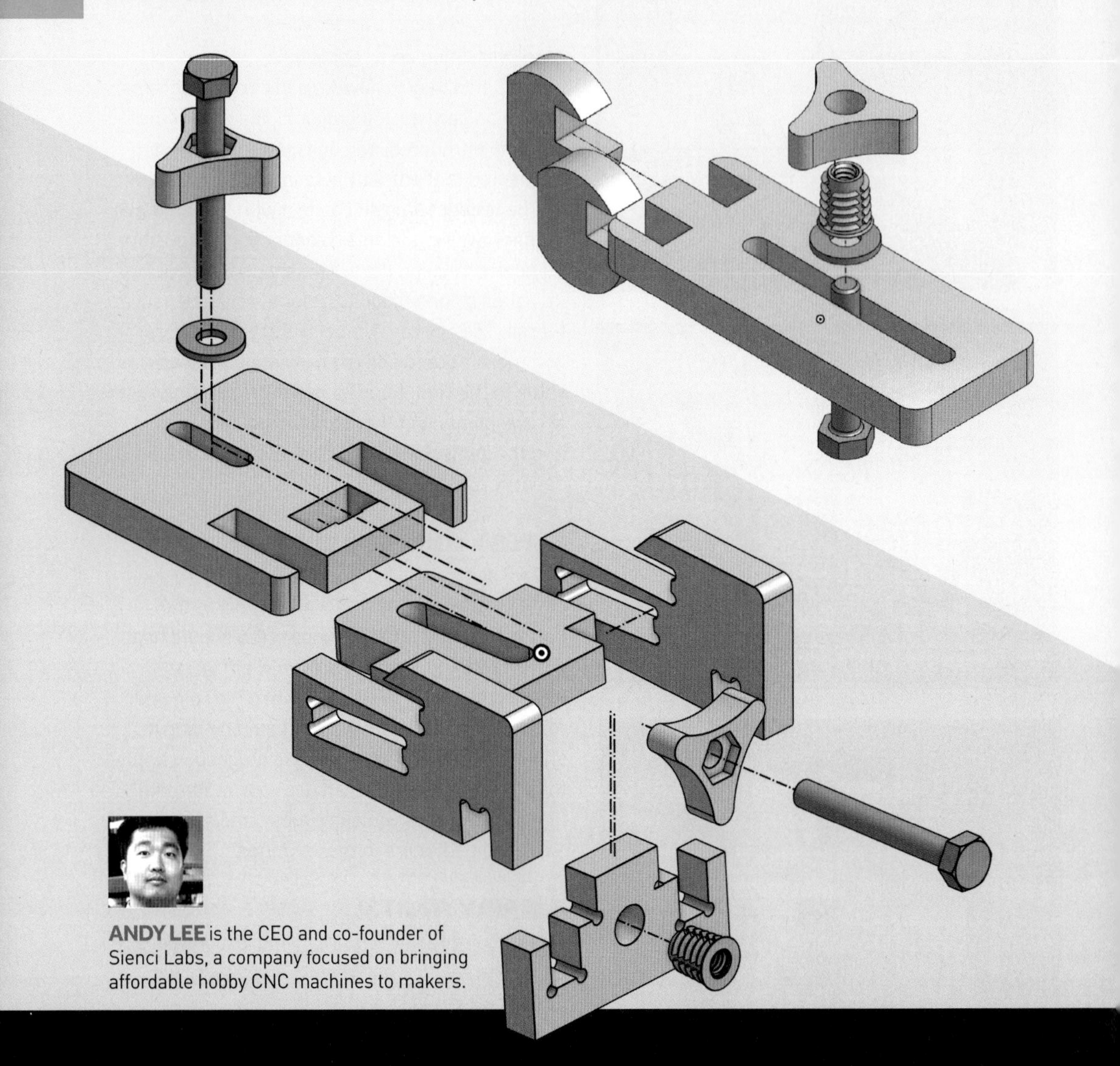

ANDY LEE is the CEO and co-founder of Sienci Labs, a company focused on bringing affordable hobby CNC machines to makers.

DIY CNC Workholding

These two universal clamps are easy to make and use on any CNC router

Written and photographed by Andy Lee

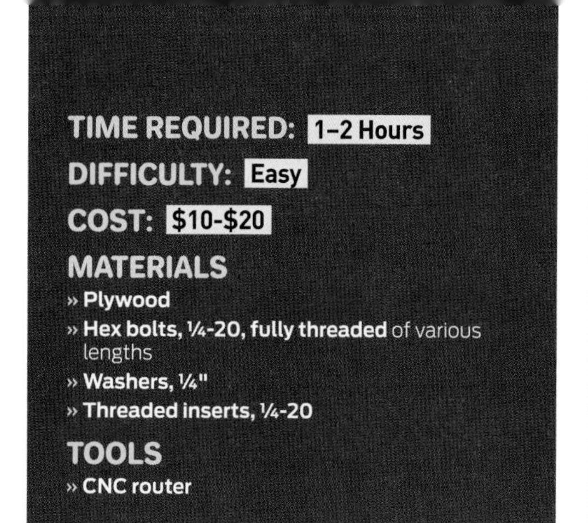

TIME REQUIRED: 1–2 Hours

DIFFICULTY: Easy

COST: $10–$20

MATERIALS

» **Plywood**

» **Hex bolts, ¼-20, fully threaded** of various lengths

» **Washers, ¼"**

» **Threaded inserts, ¼-20**

TOOLS

» **CNC router**

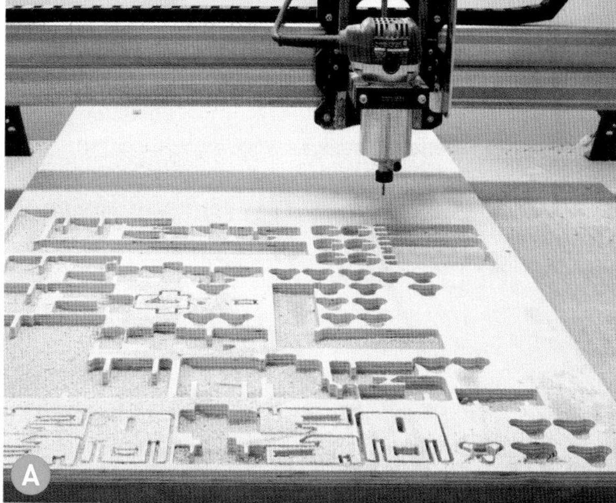

With every CNC router, strong and secure workholding is critical for a successful job. At Sienci Labs we've tried many designs, and created these universal clamps to be easy to use and make on any CNC router.

The goal for these designs is to allow folks to make their own clamps using cheap and easily available materials. For these projects, you'll need plywood, ¼-20 fully threaded hex bolts of various lengths, ¼" washers, and ¼-20 threaded inserts. Depending on the T-track you have, you may need T-bolts that fit as well, but if you're looking for a T-track for the first time, I'd recommend finding some that have a profile that fit ¼" hex bolt heads instead.

I've designed the clamps to use 11mm or ½" plywood because these are easiest to come by, but if you want to use other thicknesses, the Onshape design documents (makezine.com/go/onshape-hold-down and makezine.com/go/onshape-toe-clamp) are publicly available and can be adjusted to the materials you have on hand by changing the `#thickness` variable in Part Studio in Onshape.

GENERAL CUT SETTINGS

All of these clamps can be milled easily on a CNC machine (Figure A) and assembled by sliding the parts together. While the downloadable designs are made for 11mm and ½" material, variations in the thickness of your material may affect how well parts slide together. For the most accurate fit of parts, I recommend measuring your material's thickness with calipers, then using that thickness plus 0.1mm on the `#thickness` variable in Onshape. That being said, some gentle persuasion with a mallet will usually do the trick as well.

Everything is designed to be cut with a ⅛" bit, as that's what I've added reliefs for the radiuses on the inner corners for. Links to the DXF, sample G-code files, and other design notes can be found at makezine.com/go/two-workholding-clamps.

These are the settings I've used to cut these parts out:

- **Feed rate:** 2000mm/min on the X and Y axes, 300mm/min on the Z axis
- **Depth of cut:** 3–4mm per pass, with the last pass being set to cut 0.5mm
- **Spindle/router speed:** Around 20,000RPM
- **End mill:** ⅛" downcut 2-flute end mill

If you want to get the cleanest-looking cut, a downcut bit will work well and a compression bit will work even better.

For the sake of accessibility, these designs have been made to work with ⅛" bits. But when I make slot-together projects, I actually generally use a 1/16" corncob bit because:

1. It leaves a fairly clean top and bottom edge
2. It has a thicker overall body, which makes it less prone to breaking than a fluted 1/16" bit
3. It leaves most of the dust in the cutting path, so most of the time I can get away without needing tabs or other methods to keep the piece from flying out
4. Since the radius is pretty small, a relief on the inner corner radius isn't necessary for parts to fit together, and
5. The cuts are thinner, so it also makes less dust and waste overall.

Sienci Labs

Since 1/8" straight bits are almost ubiquitous, I've just made the designs work with those, but if you can get some 1/16" corncob bits to experiment with, I highly recommend it.

HOLD-DOWN CLAMPS

Hold-down clamps are versatile and simple to use (Figures B and C). This design is unique as it uses a rounded support at the back to allow for the right angle to apply downward pressure against your material. The optimal angle for securing your material is at a level or slightly angled down position. Based on the thickness of your material, simply flip the clamp upside down to use the side that offers the most optimal angle.

Since these clamps are made of wood, even if you have a bit of an "oops" and run into them while carving, you'll minimize the damage to the machine, and since you can make them on your CNC, you'll basically have an unlimited supply!

TIPS:

- Threaded inserts are super handy in adding threaded holes to wood. Simply fit a hex head driver or Allen wrench into the top end of the insert and screw it into the pilot hole.
- The knobs and the semicircles are prone to flying out after cutting, so I recommend milling them a little slower on the final pass than you would on the body of the clamp.
- You'll need different length bolts to accommodate different thicknesses of your project workpiece, but I've found that 1½" and 2" bolts are suitable for most applications.
- If you make the toe clamps in the next part of the article, make sure to make extra knobs as you'll need them there too!

TOE CLAMPS

If you don't want to have clamps in the way of the top surface of your material, toe clamps are the way to go (Figures D and E). By pushing in from the side, they stay away from the top of the material, and by angling the force downward, they're able to keep the material from lifting up as well.

This clamp must have a hard stop for the other side of the material to butt up against. I've also included some designs for corner stops that can be bolted to a T-track table, but any solid stop for the material will work fine.

TIPS:

- If your clamp can't get close enough to your material, try using scrap blocks to fill the gap (Figure F). This can also help if your clamps are getting in the way of your spindle or router.
- Watch out that your clamp doesn't slide away

DESIGNING IN ONSHAPE

Onshape offers a free, hobby and education use license for full functionality of their program on the cloud, with the exception that all projects made on the free plan are public and searchable. This means that derivatives of these designs will also be available to the public.

To modify a design, you'll need to create an Onshape account and duplicate/copy a new version to make changes.

Most CNC users will likely want to export all of the parts as DXFs. This is a very easy process. Simply right-click the side of the model whose *planar faces* you wish to export, and select Export as DXF/DWG. Then import the vectors into your CAM software.

Visit cad.onshape.com/help/Content/exporting-files.htm for more detailed information.

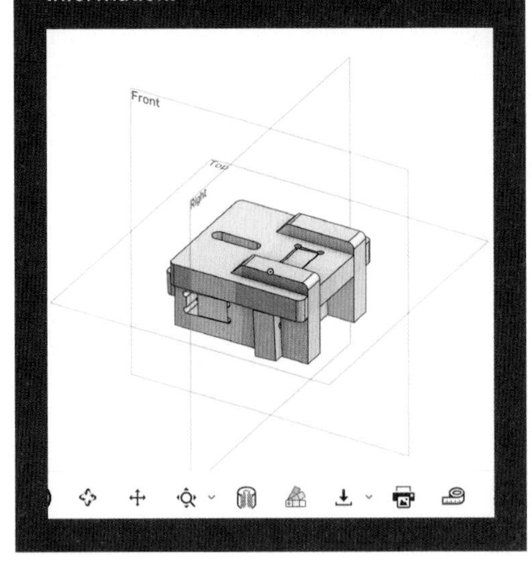

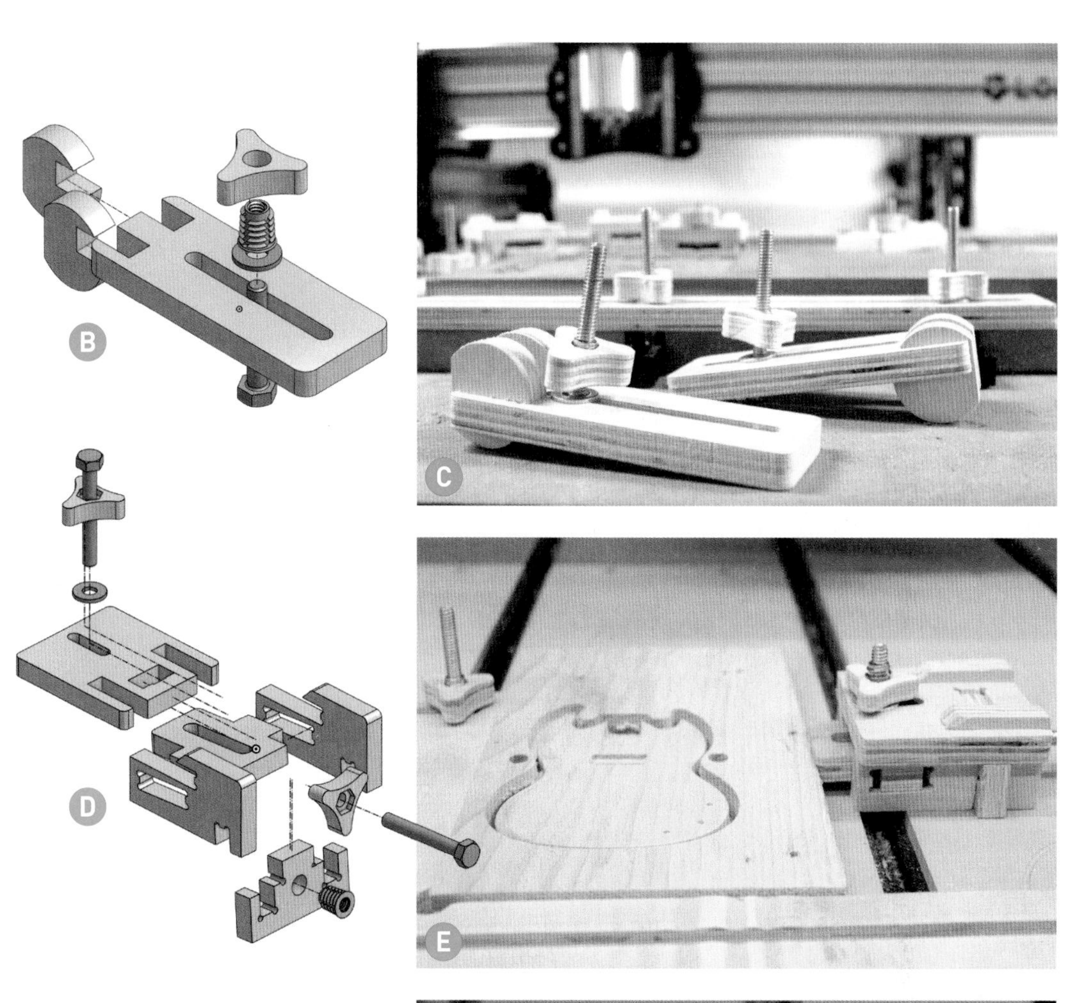

on the table when you turn the toe knob. Because of the mechanical leverage you get in the screw, the amount of force you're putting on the material may be enough to slide away the clamp as well.

FINAL THOUGHTS

I hope you find these designs useful as a starting point in building up your CNC workholding arsenal! Since these designs are freely open for you to use and modify, please feel free to make changes to the original design to make improvements and fit your needs.

Sienci Labs

The Most Annoying Noise

Build a loud, irritating — and totally unpredictable — screamer circuit

Written, illustrated, and photographed by Charles Platt

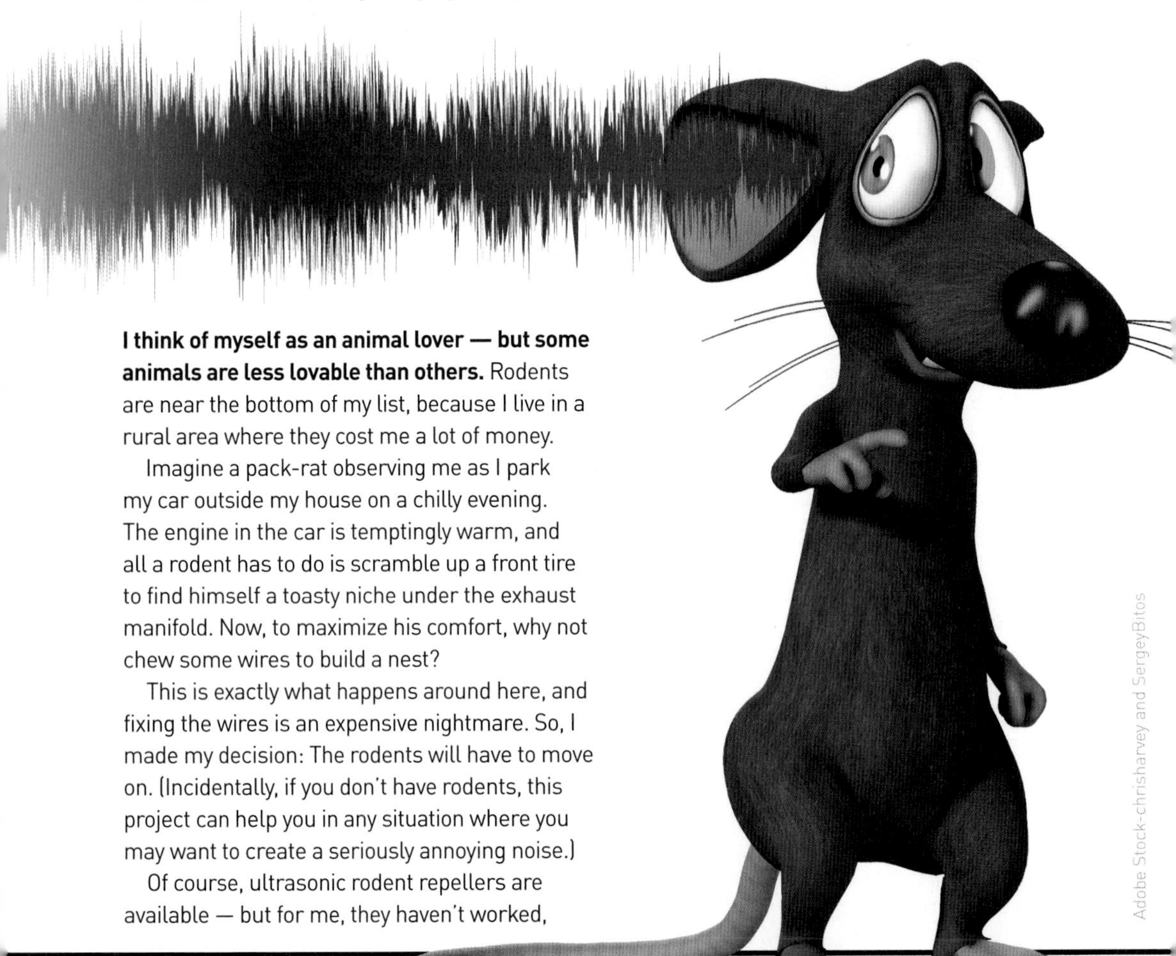
Adobe Stock-chrisharvey and SergeyBitos

I think of myself as an animal lover — but some animals are less lovable than others. Rodents are near the bottom of my list, because I live in a rural area where they cost me a lot of money.

Imagine a pack-rat observing me as I park my car outside my house on a chilly evening. The engine in the car is temptingly warm, and all a rodent has to do is scramble up a front tire to find himself a toasty niche under the exhaust manifold. Now, to maximize his comfort, why not chew some wires to build a nest?

This is exactly what happens around here, and fixing the wires is an expensive nightmare. So, I made my decision: The rodents will have to move on. (Incidentally, if you don't have rodents, this project can help you in any situation where you may want to create a seriously annoying noise.)

Of course, ultrasonic rodent repellers are available — but for me, they haven't worked,

TIME REQUIRED: 2–4 Hours

DIFFICULTY: Easy

COST: $45–$50

MATERIALS

- **Breadboard, full size**
- **Bulk jumper wire, 22 gauge (4 feet)**
- **Jumper wires with alligator clips each end (4)**
- **Stereo amplifier, 12VDC, 20 to 40 watts/channel** TPA3116-D2 or similar, from eBay
- **AC-DC adapter, 12V, rated at least 2A**
- **Piezo tweeters (2 or more)** for car stereo, eBay
- **LM7805 or L7805 voltage regulator**
- **74HC14 hex inverter chip**
- **74HC08 quad AND chip**
- **Transistor, 2N3904**
- **Standard LED, any color**
- **Basic speaker, 8Ω, 2" or larger** for testing
- **Capacitors, electrolytic, 100µF (4)**
- **Capacitors, ceramic: 33nF (2), 0.1µF (1), 0.47µF (1), and 1µF (2)**
- **Resistors: 1kΩ (1), 4.7kΩ (1), and 22kΩ (4)**
- **Resistors: 10kΩ, 15kΩ, 33kΩ, 47kΩ, 68kΩ, 100kΩ, 220kΩ, and 470kΩ (2 of each)** for experimenting

TOOLS

- **Wire cutters, wire strippers, miniature flat-blade screwdriver, needlenose pliers**
- **Multimeter**
- **Magnifying glass**
- **Capacitance tester (optional)**
- **Oscilloscope (optional)**

CHARLES PLATT is the author of the bestselling *Make: Electronics*, its sequel *Make: More Electronics*, the *Encyclopedia of Electronic Components Volumes 1–3*, *Make: Tools*, and *Make: Easy Electronics*. makershed.com/platt

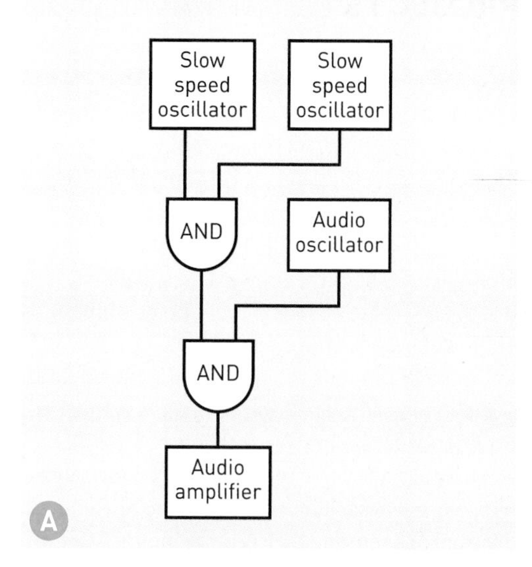

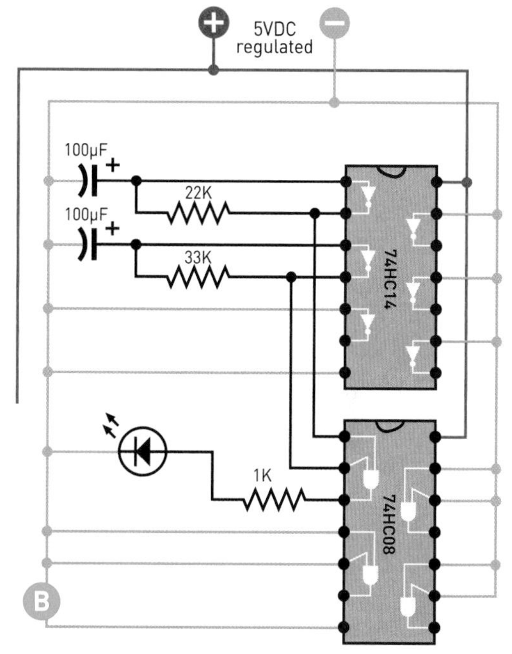

probably because they are too easy to ignore. My theory is that rodents will find a noise unpleasant if I do — so it must be extremely loud and must stop and start unpredictably. Just when a pack-rat thinks he can get some sleep, it starts again.

BUILDING AN UNPREDICTABLE SOUND SCREAMER

To make this happen, I needed a high-speed oscillator to generate an audio frequency, and slow-speed oscillators to start it and stop it. The diagram in Figure A shows what I had in mind. It may not make much sense yet, but you can refer back to it as we go along.

Because there will be multiple oscillators, I decided to use a 74HC14 chip. It contains six NOT logic gates, each of which can oscillate if you add just a resistor and a capacitor. This is much easier than using six 555 timers, or a microcontroller programmed to do six things at once.

A test schematic is shown in Figure B, using an LED to show what's going on. To breadboard it, the easiest way is by standing the resistors vertically so they fit between pins of the chip. You can see this in Figures C and D on the following page.

You'll need a regulated 5VDC supply, which

can be obtained by passing 9V or 12V through an LM7805 or L7805, as in Figure E. For this project, a 12V AC-DC converter is ideal, rated for at least 2A, as you'll need it to power an audio amplifier. Low-cost 12V supplies are easily found online.

Apply power, and the LED should flash in an unpredictable pattern. The next step will be to use this to create an intermittent audio output; but first, how does the circuit work?

Figure F reminds you about logic-gate inputs and outputs, although if you've had the wonderful experience of reading my book *Make: Electronics*, you may already know this. Figure G shows what happens if two oscillators are running at slightly different speeds, and they provide inputs A and B to an AND gate, which (as you might imagine) will AND them.

As for the NOT gates, how do they work? Turn back to Figure B, look at the first NOT, and imagine you have just applied power. The input starts low, so the output is high, because a NOT output is always opposite to its input. Current flows through the resistor and gradually charges the capacitor. The NOT monitors this, and when the voltage rises to 3V, this is close enough to logic-high, so the NOT changes its output to logic-low. The low state will sink current, so now the capacitor discharges back through the resistor into the bottom end of the NOT gate, and the cycle repeats at a speed set by the values of the resistor and the capacitor. For the test circuit, I used relatively low resistor values to create a quick result that is easy to see.

Each input of a 74HC14 has a Schmitt trigger, which creates a gap between the logic-high level and the logic-low level. In Figure H, you can see how the voltage oscillates between these limits. The red oscilloscope trace shows the voltage on the capacitor, while the green trace shows the NOT output. Figure I shows the second oscillator, which runs a bit more slowly, and Figure J shows the oscillator outputs ANDed together.

For the final version of the circuit, you'll want it to run more slowly, to give weary rodents time to nod off before being woken up again. Using a resistor value of 470K in one of the oscillators, I obtained the result in Figure K. The capacitor now takes much longer to charge than discharge, because the NOT input steals a tiny amount of

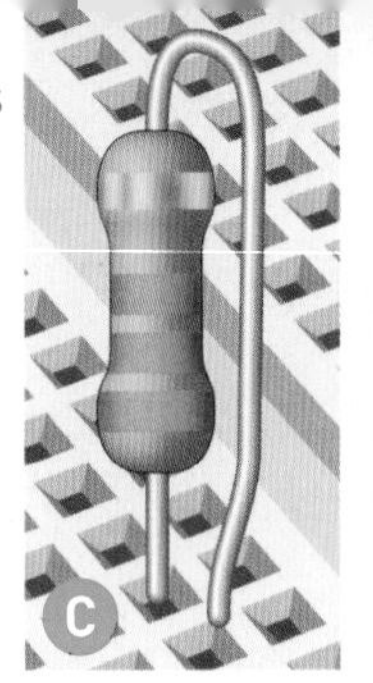

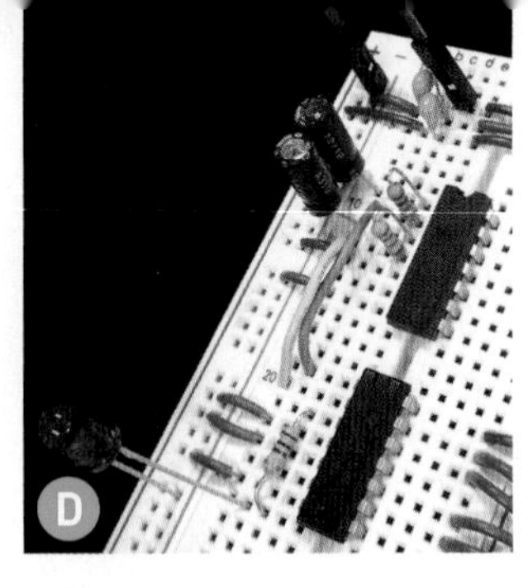

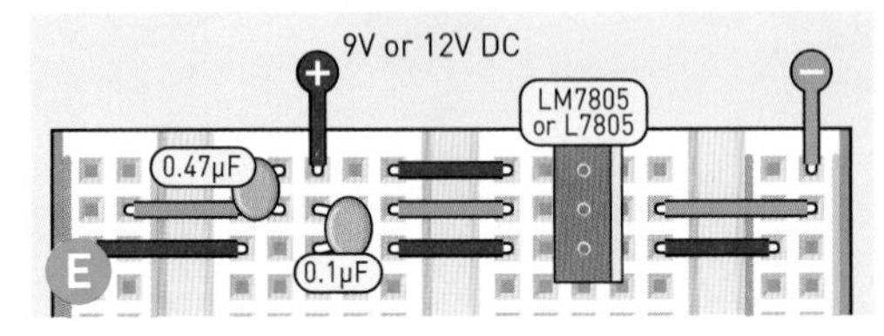

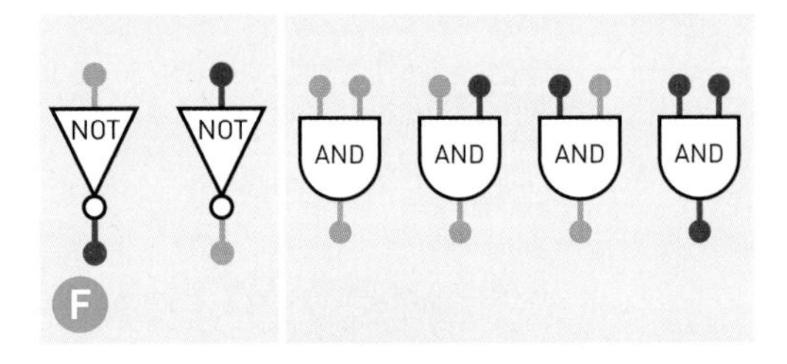

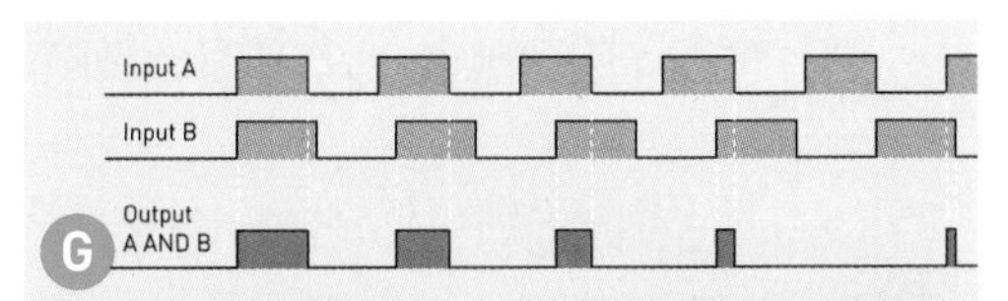

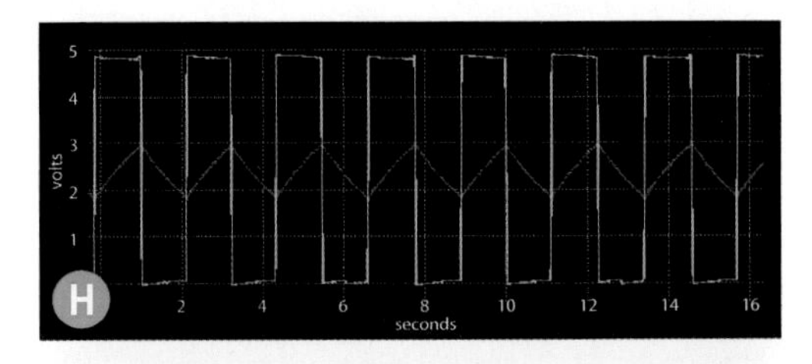

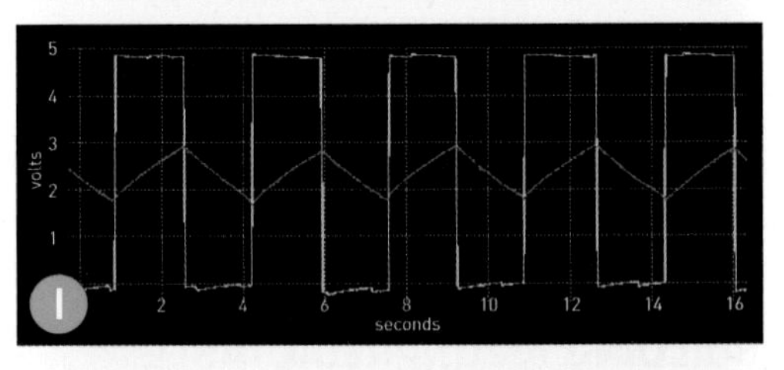

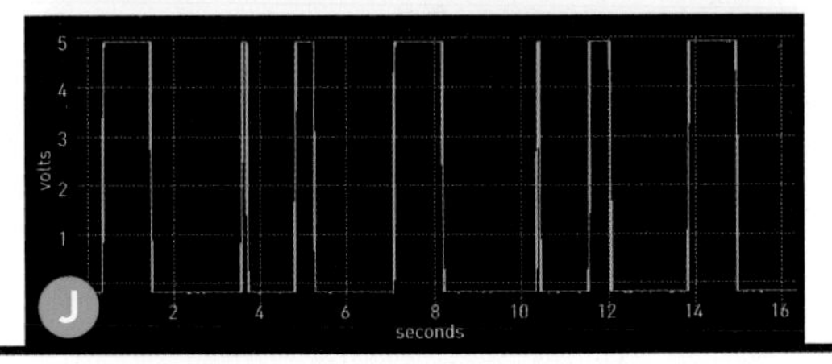

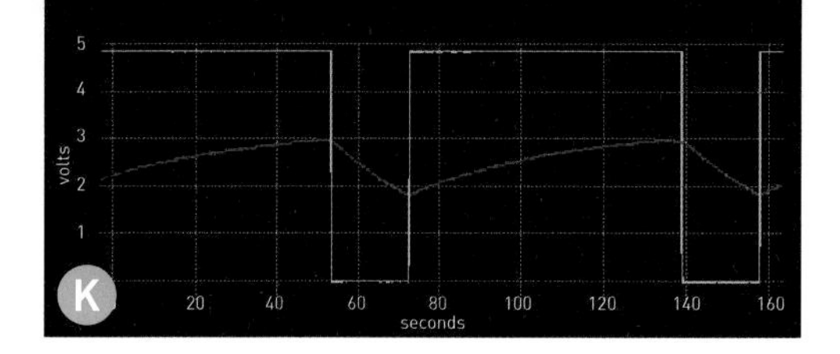

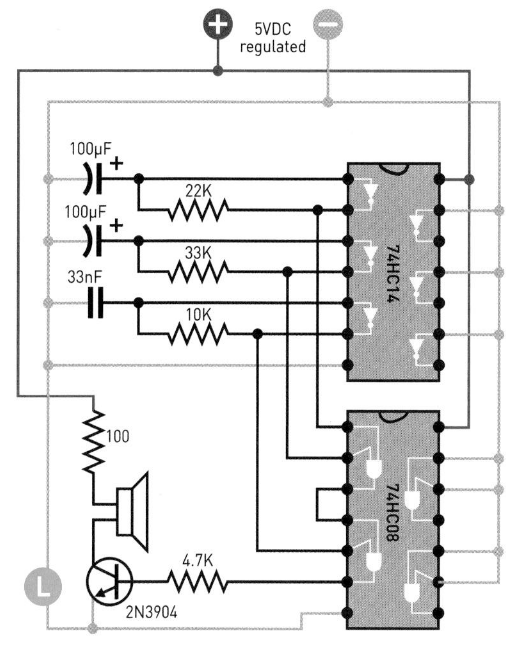

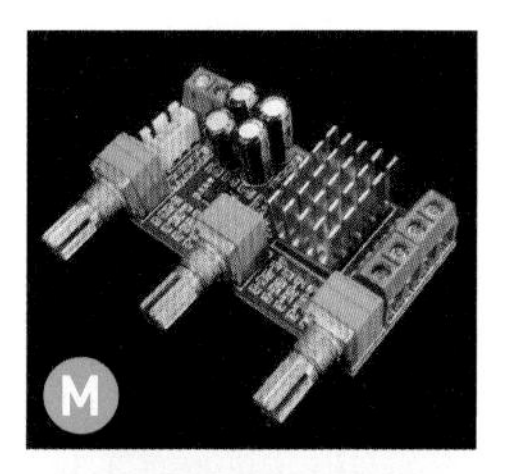

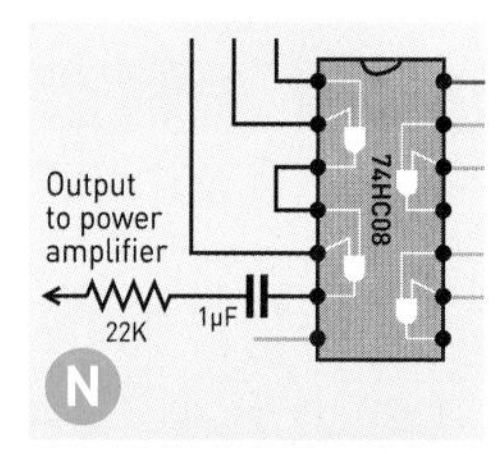

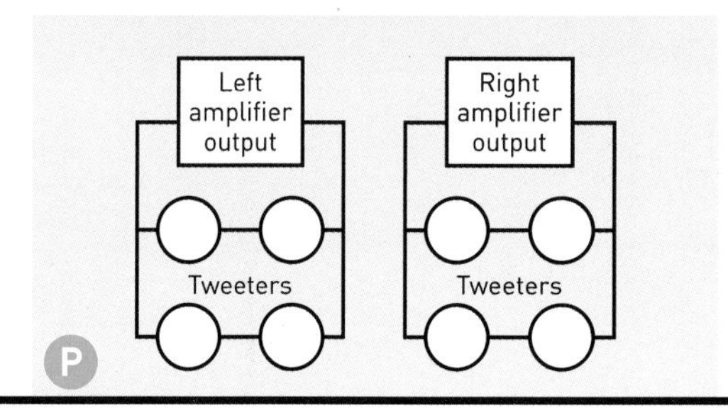

current, and the capacitor also loses some voltage through internal leakage. I think 470K and 100µF are near-maximum values; if you exceed them, the capacitor may never charge enough to trigger the oscillation. You can test this for yourself.

Now, the audio. A third NOT gate can easily oscillate at an audio frequency, and Figure L shows the test circuit extended to achieve this. The audio is ANDed with the erratic output from the previous AND gate, and for demo purposes, drives an 8Ω speaker through a single transistor.

Using the 33nF capacitor, the sound is around 3kHz. The pitch will rise if you use a smaller capacitor, but a square wave has so many harmonics, I think you'll find that it already sounds irritating.

Is it irritating enough? To do a thorough job, I decided to deliver the punishment in stereo. You can simply duplicate the circuit using the remaining NOTs and ANDs, while tinkering with the component values so that everything remains annoyingly out of phase.

AMPING YOUR RAT BLASTER

Of course, you need more power than you can get from a single transistor, so the last step is to buy a baby amplifier such as the one in Figure M, built around a TPA3116D2 chip. This can tap into your 12V AC-DC converter, or a car battery. Disconnect your test transistor and speaker, and use the resistor-capacitor combination shown in Figure N to connect with each input on the amplifier, delivering a swing of around 2V, which will keep the amp happy.

For serious high-frequency speakers, I picked up the piezo tweeters in Figure O. They're very loud, and very cheap — in fact, so cheap, I bought four pairs. You can wire them to each channel of the amplifier in series-parallel, as in Figure P, ready to bombard invading rodents with high-frequency sound from all sides.

SO FAR, SO GOOD!

No doubt, you are wondering — does it work?

Well, at this point, no more wires have been eaten. But the big test will be in the fall, when chilly nights make rampaging rodents more motivated to munch the insulation that they find so temptingly tasty.

Electro Etching

Written and photographed by Becky Stern

BECKY STERN has authored hundreds of DIY tutorials about everything from microcontrollers to knitting. She is an independent content creator and STEM influencer living in New York City. Previously she worked as product manager at Instructables (Autodesk), director of wearable electronics at Adafruit, and senior video producer for *Make:*. She enjoys riding on two wheels, making YouTube videos, and collecting new hobbies to share with you. beckystern.com

TIME REQUIRED: A Few Hours

DIFFICULTY: Easy

COST: $35–$65

MATERIALS

- » **Power supply, 12V 2A** Digi-Key #839-1688-ND. I replaced the connector with longer wires.
- » **Wire** same gauge as power supply wires, or thicker
- » **Alligator clip**
- » **Brass sheet** Amazon #B09KRLWTDJ
- » **Steel sheet** Amazon #B09L7T83S8
- » **Plastic container** larger than your brass
- » **Warm water and salt**
- » **Aquarium pump with tubing** Amazon #B0009YJ4N6
- » **Aerator stone** Amazon #B0002564ZI
- » **Electrical tape**

TOOLS

- » **Laser printer and iron/laminator/heat press** for toner transfer method

  — OR —
- » **Vinyl cutter and adhesive vinyl** for vinyl method

A

B

C

Etch custom designs into brass with salt, water, and electricity

Here's an easy method for etching metal with electricity. I wanted to make some brass plaques to commemorate my brother, who passed away recently. Other methods use toxic etchant chemicals, but this method uses salt water. Both methods have pros and cons, but I prefer the electric method because it's safer and the cleanup is more straightforward.

1. MAKE AND APPLY YOUR MASK

I designed the artwork in Illustrator to mimic the Café Bustelo logo (Figure A). (Really I only needed to make up an R and an N, since I could just trace the rest of the letters.) Before you etch, you'll need to transfer your artwork onto the brass, masking out everything that you don't want to etch.

There are a few methods for this step, as well. I first tried the toner transfer method of ironing on a laser printout, then soaking off the paper in water, but I couldn't get the temperature and pressure worked out well enough to successfully transfer my image. From what I read, there's a sweet spot in both temperature and pressure that results in success, and I just didn't get there.

So instead I used my vinyl cutter to make a sticker of my design and stuck it to the brass, which also necessitated a bunch of tweezer work to "weed" the sticker after cutting (Figure B). Make sure your metal is super clean before you apply your artwork, and don't touch it with your fingers after you've cleaned it.

2. PREPARE YOUR METAL

Make sure your power supply is unplugged. On the back of the brass, tape the stripped end of the positive wire so it makes good contact (Figure C). Cover the surface of the brass with more sticky vinyl or another waterproof tape, except where you want to etch.

Attach a gator clip to your negative wire, and

clip it to a piece of steel about the same size as your brass. Now that your two electrodes are ready to go, fill your plastic container with warm, salty water, about 1 part salt to 5 parts water. I used a container intended for storing cereal.

Place the aquarium aerator in the middle, which will keep the liquid moving and also help separate the two electrodes.

Now place the electrodes into the solution, one on either side of the aerator. The two electrodes should not touch. You can use a clamp to hold one or both to the side of the container (Figure D). You don't want to touch the bare metal once the power is plugged in.

3. ETCH IT

Now plug in the aquarium pump and the 12V power supply. As the reaction begins, you should see bubbles start to form on the exposed metal surfaces (Figure E).

I let each of my plaques etch for about 2 hours. You could speed this up by substituting vinegar for the water, but I didn't want to make a big smell.

For safety, you shouldn't leave your electro etching rig unattended while it's on. Also, this setup can create small amounts of harmful gas, depending on which metals you're working with, so do this in a well-ventilated area.

So, what's happening here? Electrical etching

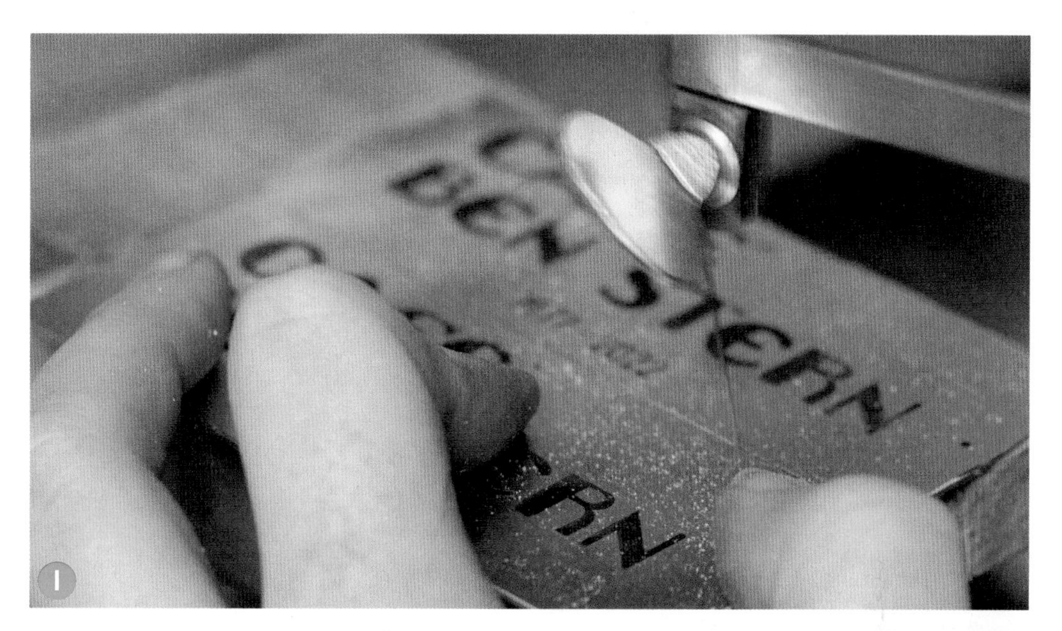

is a type of forced corrosion (Figure F). In the positively charged plate, the charge excites the electrons in the copper and zinc atoms making up the brass, and some of these atoms give up their outer shell electrons, to create positively charged ions. These dissolve in the liquid electrolyte, breaking off from the surface. Eventually, you can start to see the build-up of copper in the solution, as the etching process "bites" into the brass.

4. CLEAN IT UP

Before removing the sticky vinyl, I applied some blackening solution to provide contrast to the lettering (Figure G). Then I peeled away the vinyl to reveal the final design (Figure H). The blackening solution got under the sticker in a few places, but a quick scuff with a high grit sanding pad cleaned it up quickly.

I cut out the smaller plaques with my jeweler's saw and cleaned them up with my belt sander before drilling the mounting holes (Figures I and J).

Then I used rivets to attach one to the coffee can, my brother's preferred vessel (Figure K). He would have loved this.

To keep up with what I'm working on, follow me @bekathwia on YouTube, Instagram, Twitter, and at beckystern.com.

A

Professor Alexander Fleming at work in his laboratory at St. Mary's Hospital, London, where he discovered penicillin in 1928.

Oops! There It Is

Serendipity is the origin of many inventions and discoveries

Written by Forrest M. Mims III

If you're like most makers, you've probably made many changes and even mistakes while working on a new project. And you've probably learned so much from those unplanned happenings that your project was significantly improved. If this seems familiar, then you have experienced serendipity, one of the most powerful invention and discovery tools.

Serendipity is defined by the Macmillan Dictionary as "the fact of finding pleasant or useful things by chance." To this could be added that serendipity is responsible for some of history's greatest discoveries.

Did our ancient ancestors learn how to make fire accidentally? How did they discover that flint and obsidian can be fashioned into spear points, scrapers, and knife blades? We'll never know, but we do know about many examples of recent serendipitous discoveries and inventions, including Velcro hook-and-loop fasteners, Silly Putty, and Popsicles. Here are some others:

PENICILLIN

The antibiotic pharmaceutical penicillin came from an accidental discovery by the Scottish bacteriologist Alexander Fleming in his laboratory in 1928 (Figure A).

Fleming noticed that colonies of *Staphylococcus aureus* bacteria he was studying failed to grow in the presence of *Penicillium notatum* mold that had accidentally contaminated his experiment. He then cultured the mold and quickly verified that it produced a compound that killed not just staph bacteria but various other infection-causing bacteria as well.

Fleming received a 1945 Nobel Prize for his discovery. Many diseases have been cured and lives saved by what began as his serendipitous observation.

MICROWAVE OVEN

Percy Spencer was a brilliant, self-taught electrical engineer who received some 300 patents over his career. During World War II, Spencer developed a simplified version of the magnetron used in military radars while he worked for Raytheon.

That highly significant and carefully designed development was accompanied by a serendipitous discovery that might have cooked your latest meal: the microwave oven.

One day while Spencer was standing next to an operating magnetron, he noticed that a candy bar in his pocket had melted. When he placed some popcorn kernels by the magnetron, they began to pop. When he placed an egg near the magnetron, it exploded (invention.si.edu/node/1145/p/431).

After considerable research, Raytheon developed the Amana Microwave (Figure B), one of which I purchased for my wife Minnie in 1977. It lasted more than a decade.

Adobe Stock-caifas, Smithsonian/Martha Goodway

TIME REQUIRED:

Serendipity can occur at any time while building a project or conducting a study.

COST:

Serendipity is free. Implementing changes suggested by serendipity may increase or reduce the cost of a project.

TOOLS

» **Camera and notebooks** Besides your project tools, keep these handy to record progress — and any serendipitous findings.

FORREST M. MIMS III is an amateur scientist and Rolex Award winner. He was named by *Discover* magazine as one of the "50 Best Brains in Science." He has measured sunlight and the atmosphere since 1988. forrestmims.org

This 1967 Model RR-1 Amana Radar Range, donated to the Smithsonian Institution by Martha Goodway, was the first successful microwave oven.

POST-IT NOTES

Art Fry, a product developer at 3M, enjoyed singing in his church's choir. During a 1974 service, the paper slip he used to mark the music fell from his hymnal, and he began thinking about ways to mark the pages without damaging them.

As Fry recalled for the Smithsonian's Lemelson Center for the Study of Invention and Innovation (invention.si.edu/art-fry-post-it-note-inventor): "My mind was wandering back to the music problem when I had one of those 'flashes of insight.' Eureka! By using a recently invented adhesive, I could make a bookmark that could be stuck on, and removed, without damaging the book."

Art Fry invented Post-It Notes — using an adhesive accidentally invented by 3M colleague Spencer Silver.

The invention Fry remembered was a weak adhesive developed several years earlier by his 3M colleague Spencer Silver. When Fry applied some of Silver's adhesive to slips of paper, they instantly stuck to bare paper and could be easily removed without causing damage. The Post-it Note had arrived (Figure C). Today more than 1,000 Post-it products are sold in 150 countries.

MIMS FAMILY SERENDIPITY

Makers often experience serendipity as they develop their projects. My three children did many award-winning science fair projects during their school years. In every case, their best projects were enhanced by or even based on serendipity.

Eric wanted to build a new kind of seismometer that used a weight suspended from an optical fiber to detect earthquakes, but a geologist he consulted expressed skepticism. Why? Our Texas house is built over soil and not rock. I told Eric that meant his idea would work, for our house has a concrete foundation.

After Eric bolted his optical fiber seismometer to his bedroom floor and connected it to his computer, he began detecting seismic events ranging from heavy trucks and trains a mile away to earthquakes thousands of miles away.

His major success was serendipitous, for he also detected two underground nuclear tests in Nevada from his Texas bedroom! The judges at the Alamo Area Science and Engineering Fair in San Antonio were so impressed he received a record number of awards.

Vicki wanted to use a Geiger counter to detect solar flares. When a NASA scientist told her that wouldn't work, I told her that meant it might. Vicki conducted her project in 1989, a year with record solar activity, and she detected a dozen X-class solar x-ray flares. Her project became a chapter in a book about observing the sun.

Sarah earned many science fair awards for her research on urban heat islands across Texas, and Saharan dust over the field by our house. While attempting to use 3M Petrifilm to detect microbes in Asian dust arriving at our field, she found many mold spores and bacteria when the wind was from the south, not the west. Dust expert Prof. Tom Gill told Sarah her finding was not from Asian dust that occasionally blows across Texas. He explained that satellite images showed the likely source was biomass smoke arriving across the Gulf of Mexico from Yucatan.

Sarah tested the possibility that living bacteria and spores could be found in smoke by burning dry grass in a steel trash can. The Petrifilms she suspended in the smoke (Figure D) were loaded with colonies of bacteria and fungi.

I twice drove Sarah to the Gulf Coast, where she flew a microscope slide mounted inside a plastic cup suspended from a kite (Figures E and F). On a day when there was considerable biomass burning in Yucatan and the wind was from the south, the microscope slide was coated with numerous smoke particles and spores. On a day when there was no burning in Yucatan, only a few smoke particles and spores were on the slide.

Sarah's discovery became an award-winning science fair project ("Smoke Bugs") and a peer-reviewed, "Fast Track" paper in Atmospheric Environment, doi:10.1016/j.atmosenv.2003.10.043 (patarnott.com/atms360/pdf_atms360/04034Mims.pdf). NASA posted an article about her discovery (nasa.gov/vision/earth/everydaylife/Smoking_Surprise.html). The Smithsonian Museum of Natural History displayed a poster about Sarah's discovery, and the Smithsonian Environmental Research Center posted an article about it (forces.si.edu/

Sarah Mims suspends Petrifilms over burning grass to prove that biomass smoke contains viable bacteria and fungi.

Sarah flying a kite from Padre Island, Texas, to capture smoke particles and fungi arriving across the Gulf of Mexico from Yucatan.

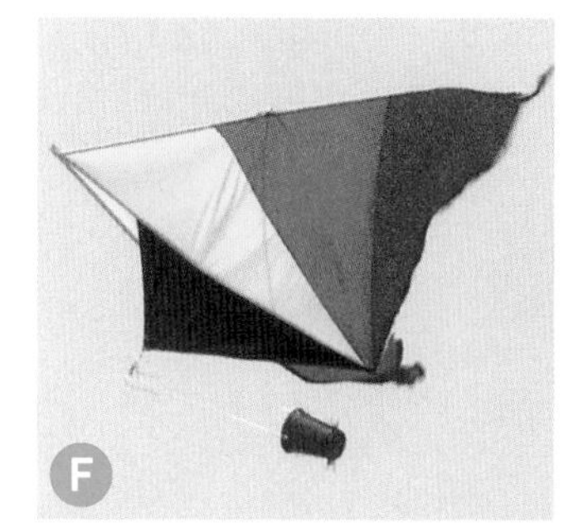

Sarah's smoke detector was a microscope slide clipped inside a plastic cup suspended from her kite.

Forrest at Mauna Loa Observatory in 1992 with his homemade TOPS-1 ozone instrument that found an error in NASA's ozone satellite.

Green ring around the sun, Dec. 15, 2022, possibly associated with aerosol particles from the historic Hunga Tonga eruption of Jan. 2022.

atmosphere/03_00_02.html). *Popular Mechanics* magazine flew us to New York City, where they gave Sarah a $1,000 Breakthrough Award and tickets to a Broadway show for her and my wife Minnie and me.

Prof. Leda N. Kobziar of the University of Idaho learned about Sarah's discovery and began to conduct serious studies about microbes in forest fire smoke. She acknowledged Sarah's serendipitous discovery as a new field of science that she named pyroaerobiology.

Some of my own science projects were significantly influenced by serendipity. I never suspected that my homemade instruments that accurately measure the ozone layer would find an error in the ozone instrument aboard NASA's Nimbus-7 satellite. But that finding was confirmed by Dobson 83, the world standard ozone instrument, during my first visit to Hawaii's Mauna Loa Observatory (MLO) in 1992 (Figure G). This unexpected discovery became my first paper in *Nature*, one of the world's leading science journals.

The LED twilight photometer I described in these pages ("Build a Twilight Photometer to Detect Stratospheric Particles," makezine.com/projects/twilight-photometer) has led to several serendipitous findings. The most significant occurred in May 2022, when one of my photometers detected a band of water vapor 56km (183,727 ft) overhead. This was an unprecedented water vapor band in the mesosphere from the historic volcanic eruption of Hunga Tonga on January 15, 2022 (see "King of the Ring of Fire," *Make:* Volume 84). My finding was quickly noticed by several scientists, and NASA's Goddard Space Flight Center followed through with an assignment for me to develop five new twilight photometers and continue making twilight measurements.

That's what I've been doing since July 2022, and during three twilights I photographed a green halo around the sun (Figure H). (These were not green flashes, which I've photographed several times from Hawaii's Mauna Loa Observatory.) So far none of the scientists I've contacted can explain this phenomenon, which is apparently another colorful example of serendipity.

Signe Dons, Public domain, via Wikimedia Commons, Forrest Mims, Grae Roth

Rock Like an Egyptian

How did they make holes in granite, or hollow a hard-rock sarcophagus? Re-create the ancient Egyptian stone drill

Written by William Gurstelle

Nina M. Davies; Louvre/Wikimedia Commons

During the reign of the pharaohs, life in Egypt was pretty good — at least compared with nearly every other place at the time. The ancient Egyptians had ample food, lived among some of the most elaborate architecture and buildings to be found in the ancient world, and even enjoyed a fair amount of leisure time.

They also had a lot of sand. Sand blew into the cities of Egypt on hot winds from the desert in enormous quantities. Its presence was a part daily life. There was sand in people's homes, in their public buildings, and even in their food. Research done on the teeth of mummies shows that most people's teeth became quite ground down due to the sand invariably mixed into their daily bread.

Sand was also used in construction of their tombs, temples, and pyramids. Not for making concrete, as concrete was invented far later, but as an abrasive in the process of drilling holes in walls, monuments, obelisks, and vases, all made of rock. Numerous descriptions in hieroglyphic writings show craftsmen drilling holes with a stone drill using sand. In fact, it was such an important tool that stylized stone drill was itself a hieroglyphic character (Figure A).

COPPER, GRIT, AND MUSCLE

The Egyptian stone drill consists of a large wooden dowel tipped with copper pipe (copper was mined, smelted, and formed into tools as early as 5,000 years ago) and turned with a hand crank or bow. The general idea is that as the copper pipe rotated against the hard rock, abrasive particles such as quartz sand, placed around the copper, slowly scoured away at the rock, making a hole. Heavy weights such as stones, suspended from the drill shaft, kept the pressure on. The stone mason needed only to keep turning the drill's handle, spinning the abrasive grit in the ever-deepening hole, until he completed the hole. Then the cylindrical rock core could be split off and, if needed, drilling could continue.

A straightforward process certainly, but it was immensely time consuming. In limestone, the material from which the pyramids were constructed, progress was slow but steady. From my experiments, it takes a couple of hours

TIME REQUIRED: A Weekend

DIFFICULTY: Moderate

COST: $75–$80

MATERIALS

- **2×6 boards, 18" long (2)** for vertical frame pieces
- **2×6 boards, 14" long (2)** for horizontal frame pieces
- **2×4 boards, 6" long (2)** for frame feet
- **1×2 boards, 7" long (4)** for corner braces
- **2×4 scrap, about 4" long** for drill guide block
- **Flat chunk of limestone rock, about 4"×4"×1½" thick**
- **Copper pipe, 1¼" diameter, 6" long** for drill tip
- **Round wooden dowel, 1¼" diameter, 40" long** for drill shaft
- **Round wooden dowel, ½" diameter, 14" long** for crank
- **Bolt, ⅜" diameter, 3" long, with nut and 2 washers** to hold sandbags
- **Sandbags (4)**
- **Sand, 50lbs**
- **Construction screws, 3" long (1 box)**
- **Abrasive powder: quartz sand, emery, or other abrasive** I used aluminum oxide sandblasting media. A pint or so should suffice.

TOOLS

- **Drawknife or other knife** for whittling the drill shaft dowel to size
- **C-clamps (2)**
- **Hole saw, 1½" diameter** for drill shaft holes
- **Drill bit, ½"** for the crank handle
- **Driver bit** for construction screws
- **Electric drill**

The stone drill was so important to Egyptians that its hieroglyphic character came to represent *art* or *craft* itself, as seen on this funerary cone of Ptahemhat (called Ty), high priest of Ptah in Memphis during the 18th Dynasty reign of King Tut. Ptah, a creator god, was the patron deity of craftsmanship.

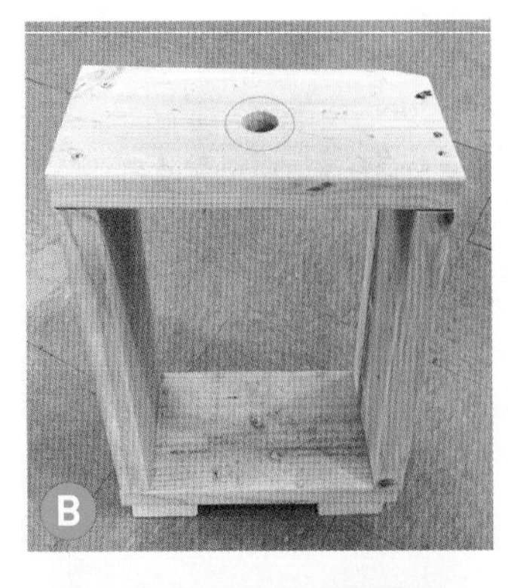

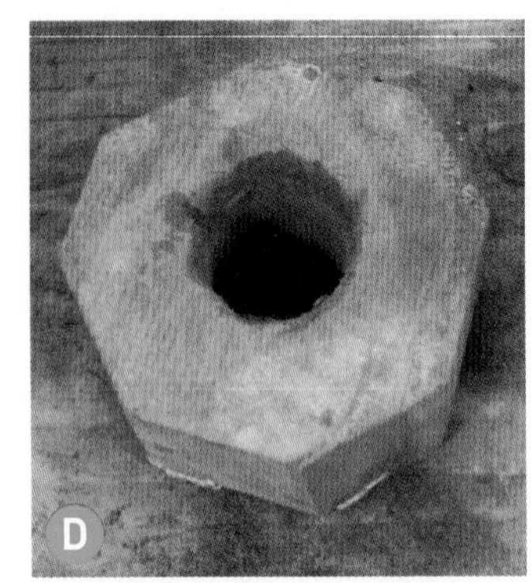

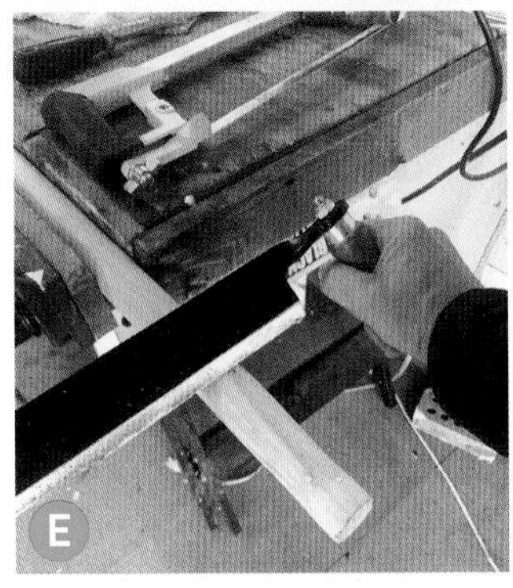

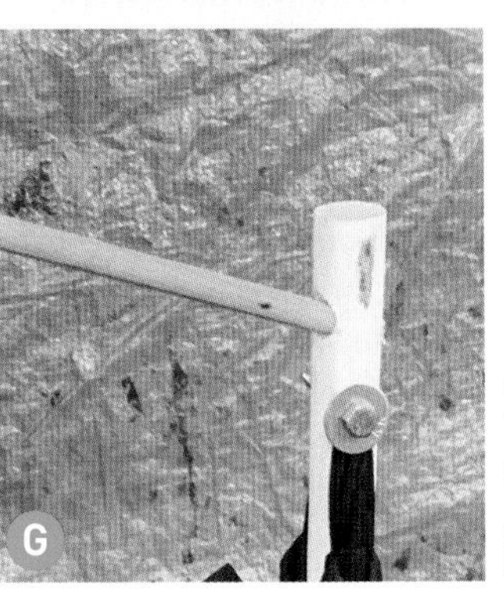

William Gurstelle

or more of constant drilling to drill through a 1"-thick rock. In harder rocks such as granite, the material from which the sarcophagi of the pharaohs were made, it takes far longer. But the Egyptians appear to have been patient workers for they drilled plenty of holes.

In this edition of Remaking History, let's re-create the ancient Egyptian stone drill.

REMAKING THE EGYPTIAN STONE DRILL

1. BUILD THE FRAME

Use the construction screws to build the drill frame out of 2×6 lumber as shown in Figure B. When the frame is finished, drill a 1½"-diameter hole in the center of the top piece as shown. Then strengthen the frame by attaching the 1×2 corner braces as shown in Figure C.

2. MAKE THE DRILL GUIDE BLOCK

Drill a 1½" hole in the center of the 2×4 drill guide block (Figure D).

3. CONNECT THE COPPER PIPE TO THE DOWEL

Use the drawknife or other whittling knife to remove slivers of wood from the round drill shaft dowel (Figure E) so that a few inches of the dowel can be inserted snugly into the copper pipe (Figure F). It's important that the copper pipe and the wooden dowel are in alignment, meaning the center lines of the pipe and the drill shaft dowel align as closely as possible.

4. FINISH THE DRILL SHAFT

As depicted in Figure G, drill a ½"-diameter hole at an angle through the drill shaft end opposite to the copper pipe. Insert the crank dowel into the hole and glue it in place.

Next, drill a hole for the ⅜" bolt through the drill shaft 2" below the crank. Insert the bolt, and then add the washers and nut as shown. You'll hang sandbags on it for weight in the final step.

5. FINAL ASSEMBLY

Fix the limestone workpiece to be drilled, with the drill guide block on top of it, to the base of the frame using the clamps, as shown in Figure H. The drill guide prevents the drill from wandering in the initial stages of drilling. (You can remove the guide after the hole is ¼"–½" deep.)

Mix the abrasive powder with some water to form a slurry (Figure I) and place a dollop of the slurry on top of the limestone inside the drill guide block. (You'll need to experiment with your choice of abrasive to find the consistency of the slurry that cuts best.)

Carefully insert the drill shaft though the hole on the top of the frame and then through the drill guide block that is clamped to the limestone block.

Add sand to the sandbags and hang them from the bolt located near the crank (Figure J). I used four sandbags weighing about 30 to 40 pounds in total as a drill weight. Once the sandbags are in place, take a moment to make sure the drill is properly balanced and won't tip over when you turn the crank.

Once that's done, you're ready to start drilling. The completed stone drills are shown in Figures K and L.

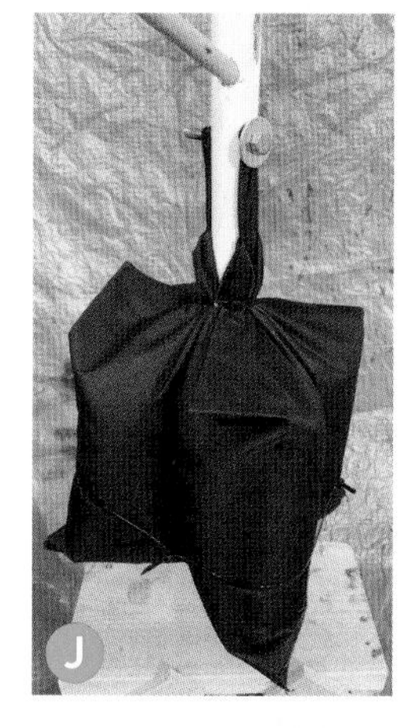

CORE WORKOUT

Grab the crank and spin the drill. As the shaft rotates, the abrasive grit between the copper pipe and the limestone will begin to cut a tubular shaped hole in the limestone (Figure M). You will be able to feel the grit cutting into the stone as you turn the drill.

When you feel the drill slide rather than cut as you turn the crank, it's time to add fresh abrasive slurry. Also, you may need to remove the drill shaft every so often to clear any packed slurry from the inside of the copper pipe.

This is not a particularly fast process, but if you are steady and attentive to the feel of the abrasive against the limestone, you will make progress.

Try to imagine yourself as an ancient Egyptian stonemason, laboring on a hot summer day, making a vase or oil lamp or even a sarcophagus for a pharaoh. You'll soon gain an intimate appreciation for how much easier we have it today with modern tools and equipment!

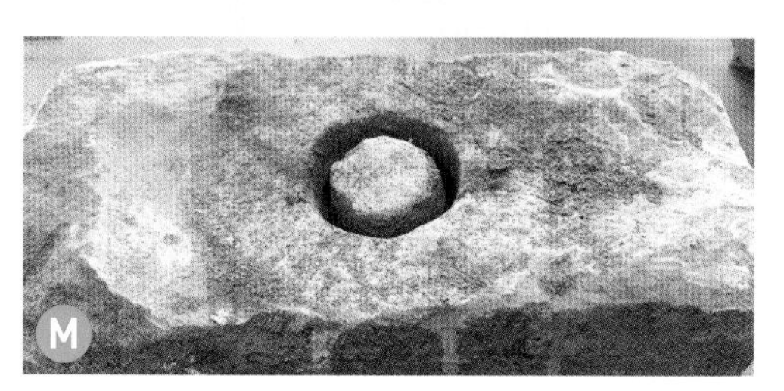

WILLIAM GURSTELLE's book series *Remaking History*, based on his *Make:* column of the same name, is available in the Maker Shed, makershed.com.

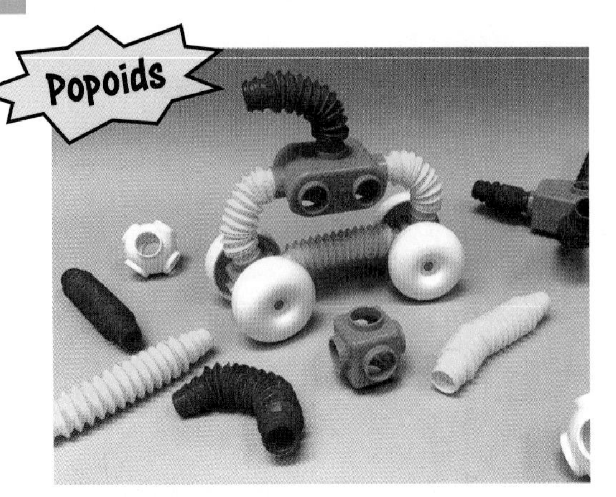

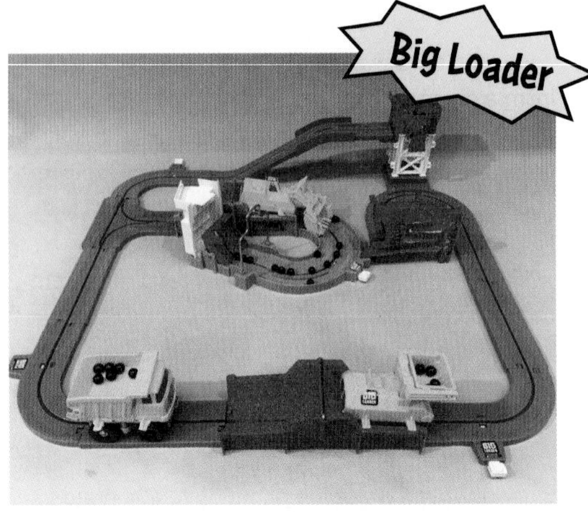

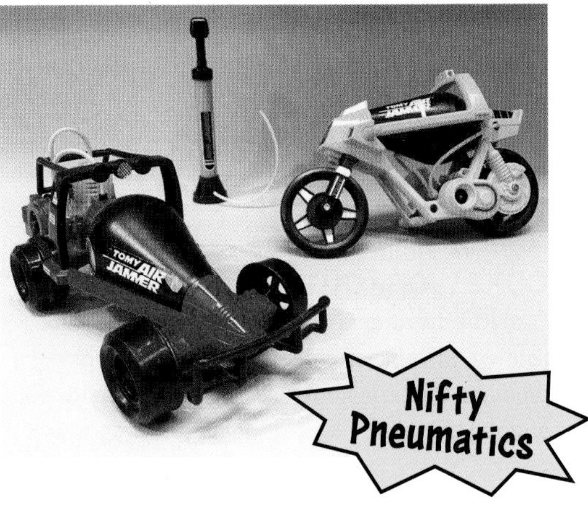

Mini Mechanical Marvels

Tomy's classic mechanical engineering toys are still amazing today

Written and photographed by Bob Knetzger

BOB KNETZGER is a designer/inventor/musician whose award-winning toys have been featured on *The Tonight Show*, *Nightline*, and *Good Morning America*. He is the author of *Make: Fun!*, available at makershed.com and fine bookstores.

As a toy inventor in the 1970s when so many toys were going electronic, I was impressed by the fantastically clever all-mechanical designs made by Tomy Toys. Here's a look at some vintage Tomy toys, with details on their still amazing gizmos.

POPOIDS

This preschool toy was simple but really fun! Stretch and pose the colorful plastic tubes into different lengths and shapes (Figure A), then join together with injection-molded parts to create figures, vehicles, or just kooky constructions (Figure B). Polyethylene tubes are flexible, but what made them also stiff? The accordion folds had sides of unequal lengths. Those alternating large/small, large/small ridges folded in a bi-stable way. The flexible wrinkles made the stiff crinkles that held their shape, like a giant elbow drinking straw. The best part was the sound and feel of the scrunchy tubes, which you can relive at youtu.be/5Aw46UoMlK8.

Today, the same fidgety, fun feeling lives on in PopSockets cell phone handles.

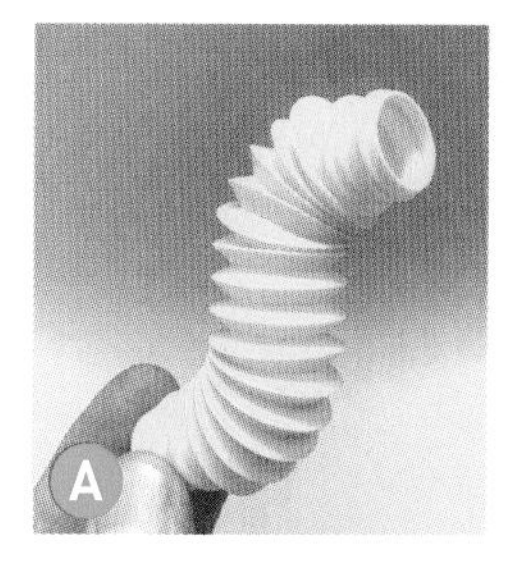

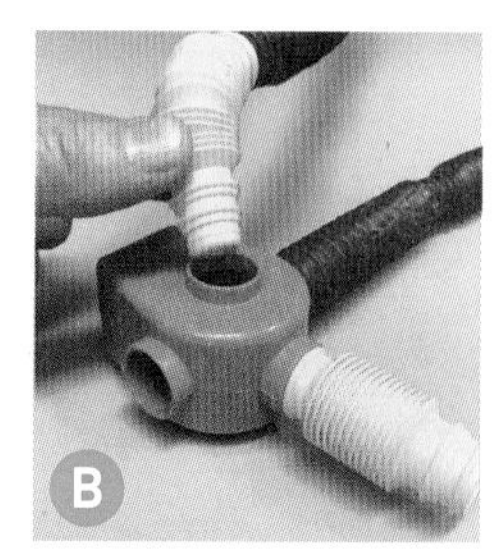

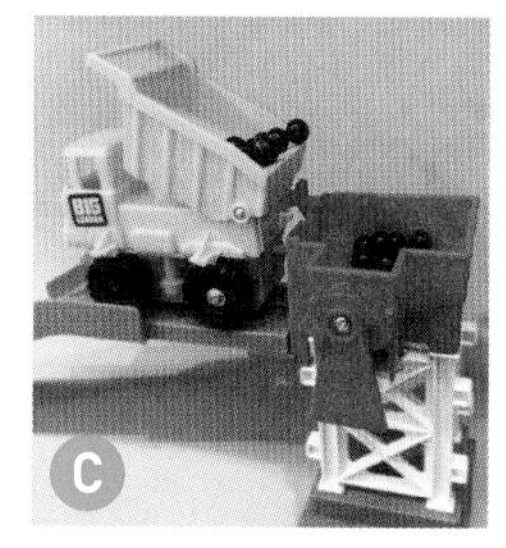

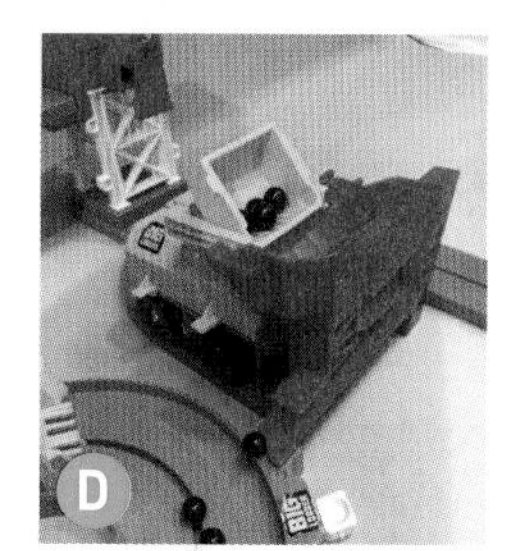

BIG LOADER

This ingenious playset has three mechanical construction site vehicles: dump truck, front-end loader, and bulldozer. They scoop, haul, and dump the plastic "rocks" (balls) into hoppers, down chutes, and around bins (Figures C, D, and E). You control the continuous action with mechanical pushbuttons to send the busy vehicles back and forth around the track. Endlessly fascinating with just the right combination of automatic action and easy control!

There's another level of fascinating action in this clever playset. All the mechanical effects are achieved with just a single motorized chassis. As it motors along, the wedge-shaped chassis sheds one body and then slides underneath to smoothly pick up a different body (Figures F and G).

Tabs and cams molded into the track activate the bodies' linkages: the truck bed raises and dumps; the dozer scoops, lifts, and dumps; and the loader catches a load and tips to unload. What makes it work is rack-and-pinion gearing. Racks are molded along portions of the track and a hidden power pinion on the bottom of the chassis

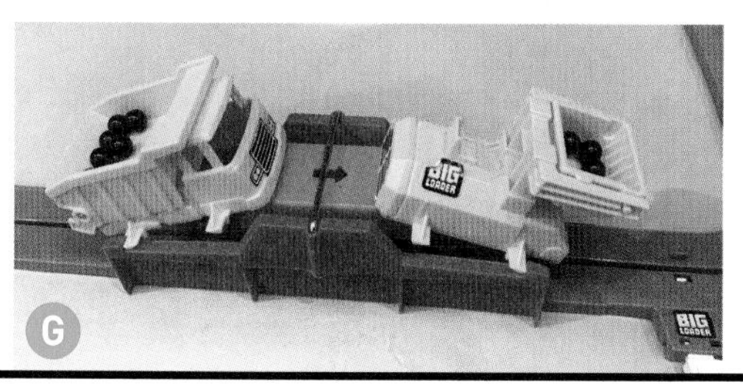

(Figure H) provides non-slip traction and extra torque to activate all the mechanisms — all from one tiny Mabuchi motor.

But what switches the direction back and forth around the track? A deeper look inside the chassis reveals the secret (Figure I). An extra pinion gear is on the rim of a tiny spinning drum (tinted pink). Indexing pegs are molded around the outside of the drum. A small paddle (tinted green) swivels back and forth to hit (or miss) the pegs to block (or allow) the drum's rotation. When the drum spins freely, the vehicle travels along in one direction, but if the paddle is swiveled over, it prevents the drum from spinning and the added gear swings into place to instantly reverse the rotation, and the chassis changes direction.

Usually changing running gears requires a bit of force as the gears are under load (which is why you momentarily let up on the throttle when shifting a car's manual transmission), but this toy's clever transmission shifts gears with just a feather-touch! Bumpers molded along the grooved track flip the hidden paddle to automatically change directions back and forth.

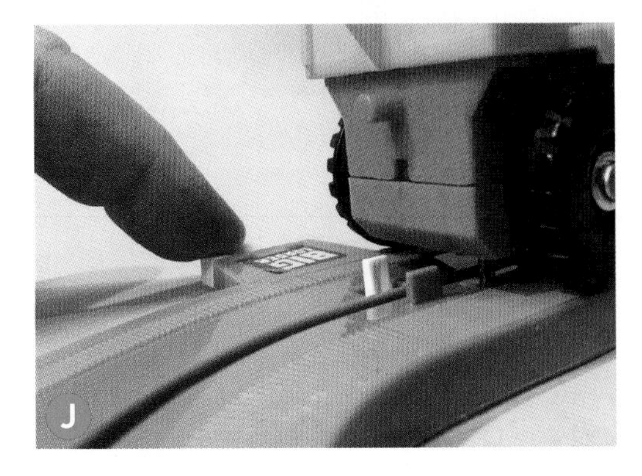

Pressing the control buttons at various locations around the track (Figure J) temporarily raises a bumper, which flips the paddle to let you change directions manually. Genius!

This amazing toy from the 1970s has been made in various versions over the years, including a deluxe set with a working elevator (also powered by the same little chassis!). The latest version is themed as a farm, with red crops replacing the black boulders, but the fun action is the same.

ARMATRON

There are lots of cool robot arm toys and kits currently available with 6-axis motion, multiple motors, and remote controls — but the 40-year-old Tomy/Radio Shack Armatron continues to amaze (Figure K).

Twin joysticks control this robot arm to grab and release objects with full rotation in the shoulder, elbow, wrist, and fingers. To make it more fun, there's also a countdown timer that cuts the power when the Energy Level bar graph reads zero (Figure L). Ingeniously, all of these multiple actions are achieved with just a single motor! A look inside reveals the marvelous mechanism.

What makes it all work is a clever system of gear trains (Figure M). Rotary motion is delivered thru the Armatron's joints using crown gears in each joint's axle (1). A four-bar linkage opens and closes the simple two-finger "hand" with rubber pads for gripping (2). The main shoulder joint provides power to all the joints via an array of nested two-sided ring gears (3), but for clarity, the illustration shows only one of the gear trains.

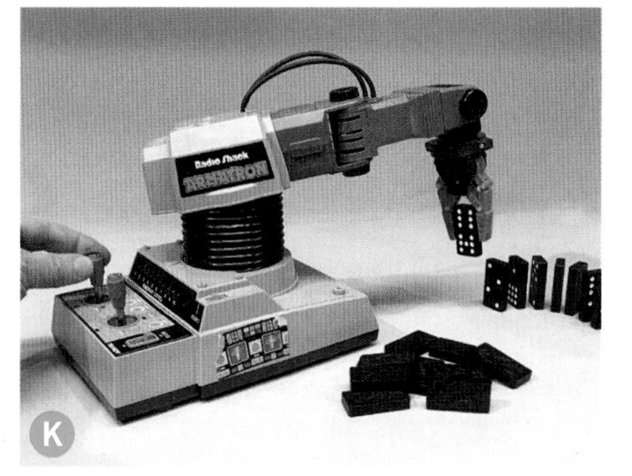

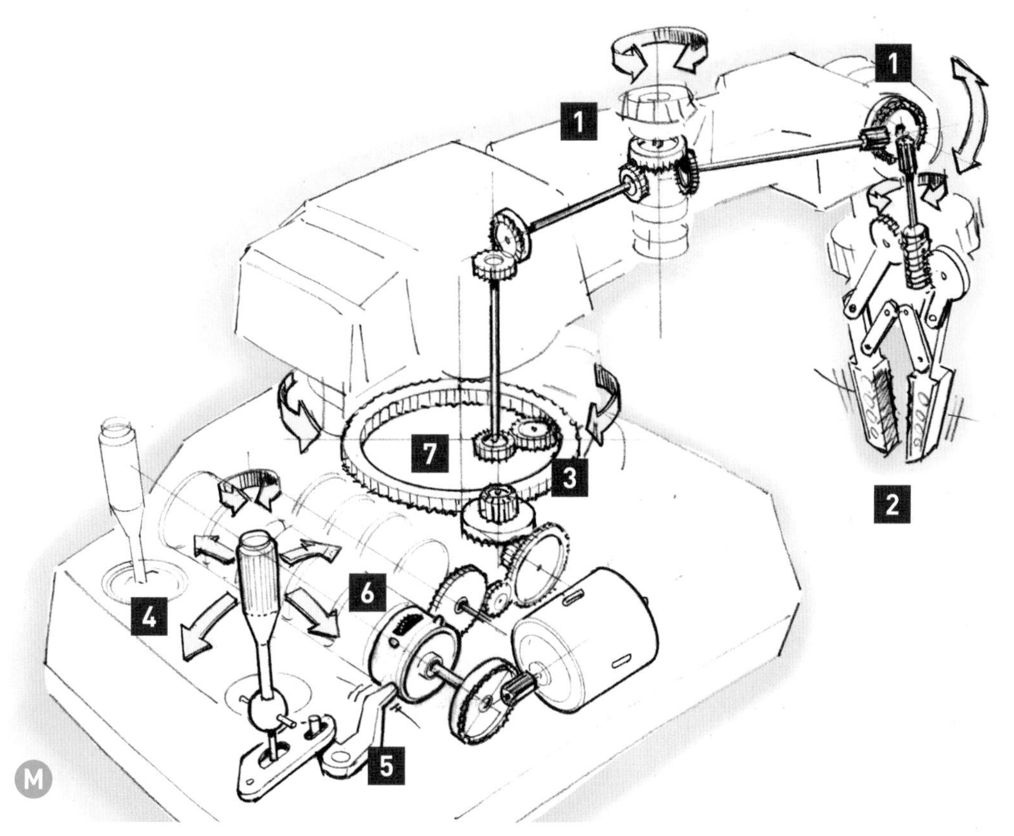

There are actually six in all! The ring gears allow a full 360-degree rotation so the arm can swing all around.

The twin joysticks (4) have mechanical linkages that translate up/down, left/right, and twist inputs to move six pointers into one of three positions: left, right, or center (5). These pointers work like the shift paddle in the Big Loader's transmission, except the design is replicated with six spinning drums (6), one for each action (shoulder rotation and lift, elbow bend, arm swivel, wrist bend, and finger grab). When at rest in the center position, the internal planetary gear spins freely and no motion is transmitted. Deflecting the joystick moves the arm left (or right) to stop the drum in one position (or the other), which swings a drive gear pinion into action. The drum then transmits torque to its linked ring gear (7) in the main shoulder joint and down through the rest of the gear train.

This patent drawing shows only a small part of the Armatron's gear trains!

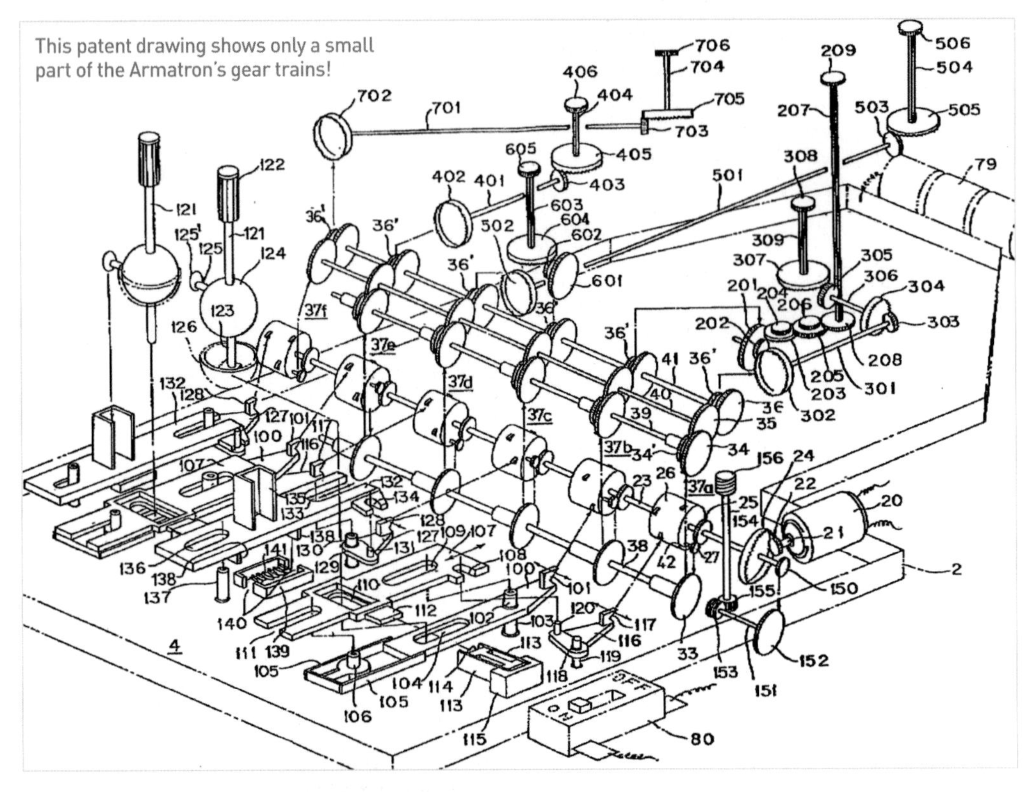

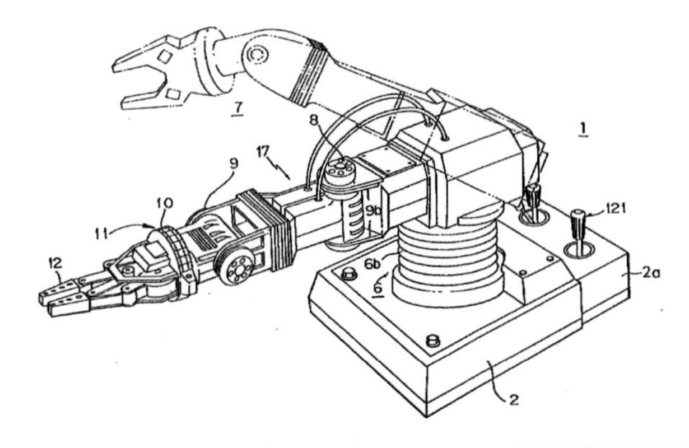

Some of the drums even have added gearing and five left/right pointer positions for slow/fast speed control. Spring-loaded clutches prevent stripping gears or stalling the motor when you bump into something (or your little brother grabs it).

At the same time, another gear train from the same motor also turns a worm gear for a big reduction in RPM to work the countdown timer. The Energy Level bar graph drum with number wheel shows how much energy (time) is left until the power to the motor is automatically cut off by a mechanical switch. Better hurry — can you complete your robot arm mission before time runs out?

Check out this YouTube video (youtu.be/zCiyR8s1vCo) to see the insides of the Armatron in action!

There is a passionate, dedicated fan base for the classic Armatron. Longtime *Make:* contributor I-Wei Huang even hacked his Armatron (Figure N) to create a Maker Faire-worthy steam-powered version! crabfu.com/steamtoys/steam_armatron

The closest toy robotic arm available today is OWI's Robotic Arm Edge kit. It uses five separate

motors on flying leads to achieve five DOF and control. Not as mechanically clever, and it's tricky to assemble, but a good lower-cost design trade-off.

NIFTY PNEUMATICS

Tomy continued to make many other unusual and cleverly engineered toys through the decades: water-filled mini skill games, all-mechanical hand-held "electronic" pong games, and wind-up toys and games. Memorably, they pioneered air-powered Air Jammer cars and cycles with pump-up pressure tanks and piston motors (Figure O) — satisfyingly noisy, safe, and fun!

Air power also activated Tomy's Space Pets, a line of mechanical creatures with names as wild as their action. High Hopping Hoomdorm had a six-legged insect-like hopping action, while Stretch Legged Stoomdorm inch-wormed its way along using an extending, scissors-like chassis (Figure P). Kooky and clever!

Imagine what Tomy could have done combining their air-powered toy tech with the latest robotic arm designs, like tentacles (youtu.be/SayuM8E_WaQ) and "jamming" soft-touch grippers (makezine.com/projects/universal-robot-gripper)!

Tomy merged with Takara in 2006. Now much of their toy business is oriented toward licensed toys from TV and movies, but Tomy's earlier mechanical marvels live on in the minds of toy inventors and curious engineers.

A Transparent vase mode.

TRANSPARENT 3D PRINTS

Written and photographed by Rich Cameron

Tips and techniques for getting glass-like prints from clear filaments

In order to create glass-like transparent 3D prints, you need to start with a very transparent filament — but there's much more to it than that. After experimenting with different filaments for the last 10 years, I can share my tips for getting great clear prints.

First, your transparent filament. Most natural PETGs will work well, though you'll likely find that some work a little better than others. The photos in this article use a mix of natural PETG from Overture and Sainsmart, and Hatchbox transparent black PLA. PLA is rarely as transparent as PETG, and higher-temperature materials like PC tend to cloud while cooling.

It doesn't matter how transparent the filament is if you have lots of interstitial spaces breaking up the light going through, the way you will with typical infill settings. This means that you either have to create a hollow print with a single wall, using vase mode for example, or your print has to be completely solid.

VASE MODE

Taulman3D, the makers of T-glase, the first PET-based 3D printing filament, gave us the recipe for creating super-transparent prints in vase mode. It turns out that in this case, extremely thick layers are best (Figure A). You want to use a layer height close to your nozzle width, so that there is minimal contact area between layers, and most of the light entering one extrusion line will exit out the other side of that same line, rather than refracting around unpredictably as some rays cross the boundary between layers.

A layer of a clear coat like XTC-3D to fill in the space between these lines further enhances the effect (Figure B).

SOLID PRINTS

Getting transparency with anything thicker than a single wall requires filling in all the tiny spaces between extrusion lines. This means you need to make your prints solid, either by using **100% infill**, or by skipping the infill altogether and making all your layers **top/bottom layers**. Filling all the space in these layers is easier with thin layers, but that's not enough.

You'll need to carefully calibrate a **flow multiplier** that works for your particular printer and filament. This will probably require turning up your flow multiplier, since most printers slightly under-extrude by default.

In order to get the plastic to flow into itself well enough to squeeze out all the air, you'll need to print **hotter than usual** to lower its viscosity. Also, you need to print **really slowly**, in the 10–30mm/s range, so that the plastic can flow into all the little cracks, and give the air time to work its way out before it gets trapped.

TOP AND BOTTOM SURFACES

To be able to see through a print, you also want to make sure that the top and bottom surfaces are as smooth as possible. For the top surface, there's a setting called ***ironing*** designed to help with that by doing a second smoothing pass over the top layer. However, ironing algorithms generally extrude a little bit while ironing, assuming that the top layer will not be completely filled. If you've adjusted your flow rate to avoid

RICH "WHOSAWHATSIS" CAMERON is an open-source 3D printer hacker who designed the RepRap Wallace and Bukito 3D printers. He is a co-founder of the Pasadena, California-based consulting and training firm Nonscriptum LLC, focusing on teaching educators and scientists how to use maker tech.

B Clear coats enhance transparency.

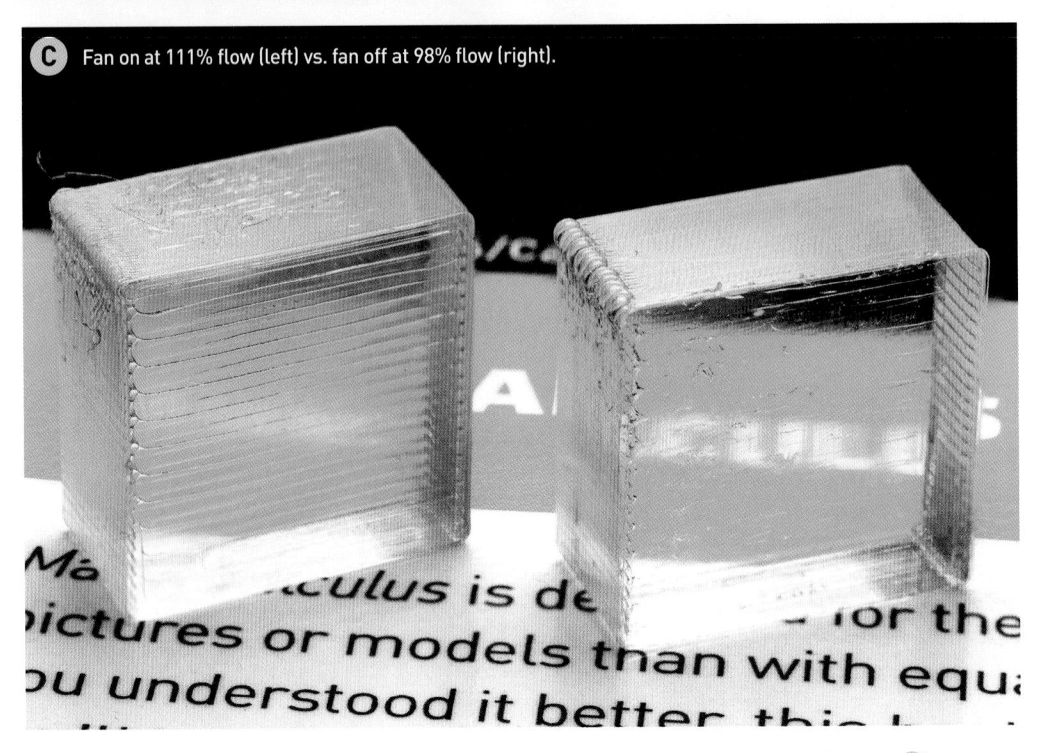

C Fan on at 111% flow (left) vs. fan off at 98% flow (right).

capturing air, this won't be the case, and you'll need to set the ironing flow to zero.

The bottom surface might be a bit trickier. In order to get a smooth, glassy bottom surface, you need to print on something with a smooth and glassy surface. Glass is the obvious choice — but PETG prints on bare glass have a tendency to stick too well and occasionally pull a razor-sharp shard of glass off with them. Using glue or hairspray as a release layer would cloud the surface. PEI print surfaces are common, but PETG also tends to stick too well to those, especially the smoother ones, and the high temperatures involved will make it worse. What I found worked best was a strip of polyester-based **high-temperature masking tape**.

PRINT COOLING

Print cooling also makes a big difference. Once the plastic leaves the nozzle, it immediately starts to cool. As it cools, it shrinks, so the more it cools between layers, the more plastic you'll need to put down to fill the space.

On one printer, I needed a flow multiplier of 111% with my cooling fans on, but I got even better results by **turning the fans off** and lowering the flow rate to 98% (Figure C).

Even with the space filled, the fan-cooled prints were more cloudy than the prints that were kept as hot as possible between layers. This means that you not only want to turn the cooling fans off, but **enclose the printer** for best results, to slow the natural cooling. This slower cooling also means that the cooling from one layer to the next will be more consistent as the area of the layer changes, which makes it more useful for printing things other than calibration cubes.

UNIDIRECTIONAL PRINTING

When I read Ryan Cooper's (Rygar1432) recent guide on Printables (printables.com/model/15310-how-to-print-glass), I was surprised by the suggestion that you should avoid printing layers in alternate directions. I had previously used alternating directions, and thought that method would make it easier to fill any small gaps in the previous layer. When I tried the **unidirectional method** though, I found that it did help, for a few reasons (Figure D). First, if any gaps remained, they lined up every layer, creating lines of non-transparency instead of a grid. Second, it meant that plastic was added to

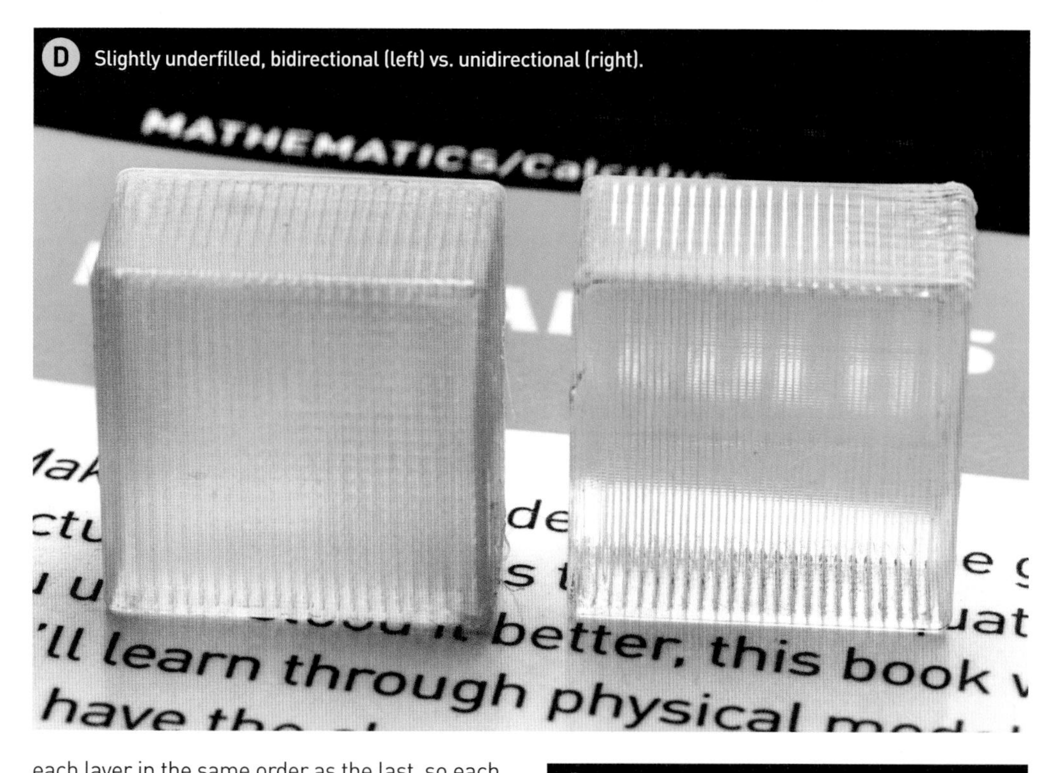

D Slightly underfilled, bidirectional (left) vs. unidirectional (right).

each layer in the same order as the last, so each part of the layer had the same amount of time to cool before the nozzle came back around to print on top of it. This made a significant difference, especially when I had the cooling fans on, and will make a bigger difference in larger prints that have more time to cool between layers.

SURFACE FINISHING

Finally, the layer lines will always create some surface texture that makes side surfaces — and sloped surfaces in particular — less transparent. As with most 3D prints, surfaces will look best either **perfectly horizontal or close to vertical**.

To get the best clarity you'll need to smooth out those layer lines, either by using a **clear coat**, or by **sanding** them smooth. The print shown in Figure **E** was wet-sanded in about 20 steps, starting at 80 grit and ending with 1-micron polishing paper (equivalent to approximately 15,000 grit).

E Solid and sanded.

HOME AUTOMATION FOR MAKERS

Open source Home Assistant can connect all the things — including Matter and DIY devices

Written and photographed by Wayne Seltzer

Makers seem to like connecting things together. Why is this? Of course, there's the convenience of remote control, which almost everyone appreciates. But there's something strangely satisfying when previously unassociated devices can interoperate in a synergistic whole — the sum is greater than the parts.

This is what I've always liked about home automation (HA) technologies, from the earliest TV "clickers" to the latest innovations like **Home Assistant** and **Matter**.

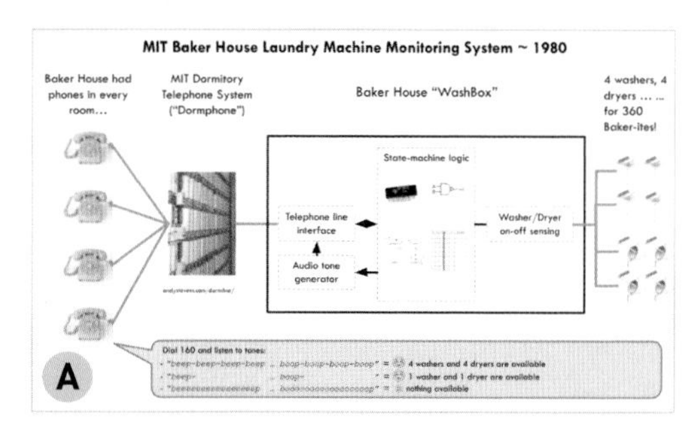

MY HOME AUTOMATION PATH

My family were early adopters of consumer electronics in the 1960s. Impressed by our Zenith TV's "Space Command" remote control, I took it apart and learned that it wasn't electronic, but acoustic! I soon discovered that shaking a bunch of keys on a ring could also control the TV. Mom was not amused. Dad tolerated my experimentation in home automation security.

Thus began my HA journey. As an MIT student I automated current-sensing coils and a phone line to report the availability of washers and dryers in the laundry room (Figure A). When I bought my first home I installed X10 switches, outlets, and controllers. And when I got my first home PC, I bought an X10-to-serial interface so I could create rules for those devices: *Turn on the porch light at sundown*.

In the early 90s I worked for a company creating prototypes for the new CEBus/EIA-600 standard, intended to enable multi-vendor home automation. Network layers included powerline carrier (like X10) but also RF, infrared, and coax. More importantly, CEBus included reliable, secure communication and an extensible language to accommodate all kinds of HA devices for lighting, security, energy management, etc.

CEBus was widely adopted — but failed. What was missing? Low-cost, small microcontrollers, and wireless networking like Wi-Fi and Bluetooth.

IoT, Wi-Fi, BLUETOOTH, ZIGBEE, Z-WAVE

By the 2000s, home Wi-Fi, internet, and smartphones enabled a new generation of home automation products accessible to a much larger market. Wireless switches, plugs, and outlets became easy to install and configure with manufacturer's apps on your phone, which then control your devices through the manufacturer's cloud services. (Easy, as long as the Wi-Fi and internet are operating perfectly.)

More products entered the market from Amazon, Apple, Google, Samsung, Philips, and many new vendors, adding more complexity and automation possibilities. New networking standards including Zigbee and Z-Wave enabled mesh networks to connect devices throughout the home, extending reach beyond Wi-Fi.

But competing standards made it challenging to create an HA system with products from different vendors. And proprietary protocols gave do-it-yourself makers limited opportunities to build their own devices that could work with a vendor's products. Sure, you could become an Apple HomeKit or Google Home developer, but this is not for the casual maker.

I gradually replaced my flaky X10 devices with light switches and outlets based on Tuya (tuya.com), an HA platform available to many vendors; a Nest thermostat; and a water-leak detection system from a discontinued Zigbee-based platform, Eaton's Home Heartbeat. My son Michael and I built a Raspberry Pi solution to integrate with Home Heartbeat, enabled by the reverse engineering documented by Blooming Labs makerspace (bloominglabs.org/index.php/Eaton_Home_Heartbeat). Fun, but it still couldn't readily integrate new HA devices and services.

HOME ASSISTANT: OPEN SOURCE TO THE RESCUE

During Covid, I stumbled onto the impressive **Home Assistant** project (home-assistant.io): "Open source home automation that puts local control and privacy first. Powered by a worldwide community of tinkerers and DIY enthusiasts.

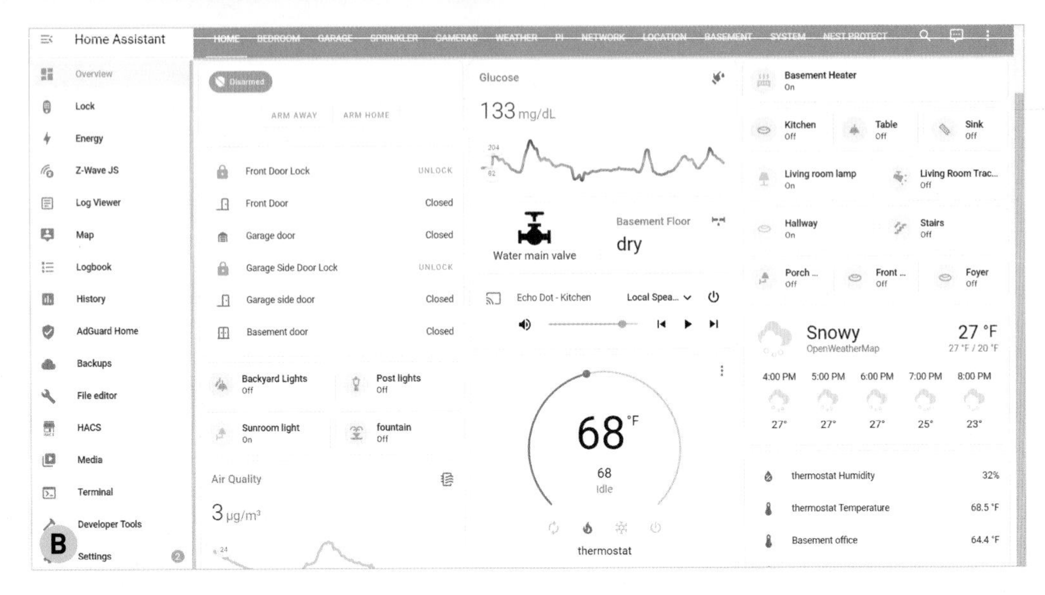

Perfect to run on a Raspberry Pi or a local server."

Instead of an HA standard, Home Assistant is a platform that allows open-source developers to create integrations for any device or service. There are over 2,000 integrations available, with over 7,000 contributors, and more than 200,000 active installations.

Home Assistant's dashboard (Figure B) is easy to create and edit with a visual editor (Figure C).

Home Assistant is very much a maker-friendly project; you download the software to a USB drive and install it on a Raspberry Pi or other Linux server, possibly in a virtual environment. It is now possible to purchase a Home Assistant appliance, ready to plug in, but it's still not something that general DIYers are purchasing from a big-box home improvement store. Maybe in the future?

Key to the success of Home Assistant are ***automations***. They have three components: ***triggers***, with optional ***conditions***, cause ***actions*** to occur. For example: When the front door is opened (trigger), and it's after sunset (condition), turn on the hallway light (action). Automations are also easy to create with a visual editor (Figure D).

Because Home Assistant has so many integrations — all the big platforms; brands like Phillips, Lutron, Leviton, and Ikea; protocols like MQTT, ESPHome, and Tasmota — there are endless ways to build your own HA device and make it part of your Home Assistant ecosystem.

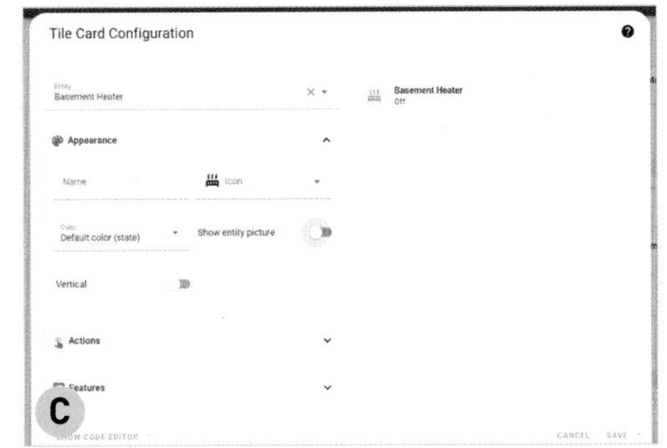

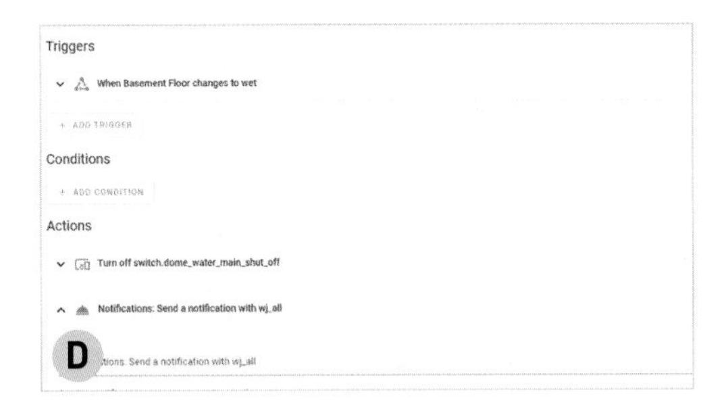

Here are some of the HA projects I've implemented with Home Assistant:

LEAK ALARM / WATER MAIN CONTROL

When you think about the damage an unnoticed water leak can cause, you can't help but consider an HA solution. I built a simple leak detector using some gold-plated traces on a surplus printed circuit board and connected it to an ESP8266 D1 mini to monitor the resistance.

The ESP8266 microcontroller is great for HA projects as it is low cost, includes Wi-Fi, and is supported by the Arduino IDE. The Arduino MQTT library works well with the Home Assistant MQTT server using a publish/subscribe model: the Arduino client publishes wet/dry MQTT events which are discovered by the MQTT server. This creates an ***entity*** in Home Assistant which can be used as a trigger in automations.

There are several water valve shutoff products that integrate with Home Assistant. I used a Z-Wave valve that's wirelessly paired with a Z-Wave USB interface on my Home Assistant Raspberry Pi.

The automation is:

- **Trigger:** Leak detector entity changes to the "wet" state.
- **Actions:** Turn off water valve, send a text message and email, turn on a piezo alarm connected to the Raspberry Pi GPIO.

HUMANE MOUSETRAP WITH TASMOTA

Small furry uninvited houseguests often meet their fate with lethal mousetraps, but they're just trying to live, right? A humane mousetrap is the better approach for some, but if you don't notice that the mouse has checked in, the poor critter might really suffer.

So, time to build a better mousetrap. I bought a humane mousetrap and added a magnet and a reed switch connected to an ESP8266. I chose Tasmota open source firmware for the ESP (tasmota.github.io), in conjunction with the Home Assistant Tasmota integration. A web server on the microcontroller, programmed in Arduino, enables configuration such as selecting Wi-Fi SSID and password, device name, and configuring the IO pins. In this project, the mousetrap detector appears as a ***binary sensor*** to Home Assistant, with on/off states (Figure **E**).

- **Trigger:** The door of the mousetrap is tripped.
- **Action:** Send a text message, and tell Amazon Alexa devices to announce "There's a mouse in the mousetrap." I soon learned that this mouse event is most likely to occur in the middle of the night and wake everyone up. Not cool!
- **Condition:** So, a simple change to the automation delayed the announcement until after our morning wake-up time.

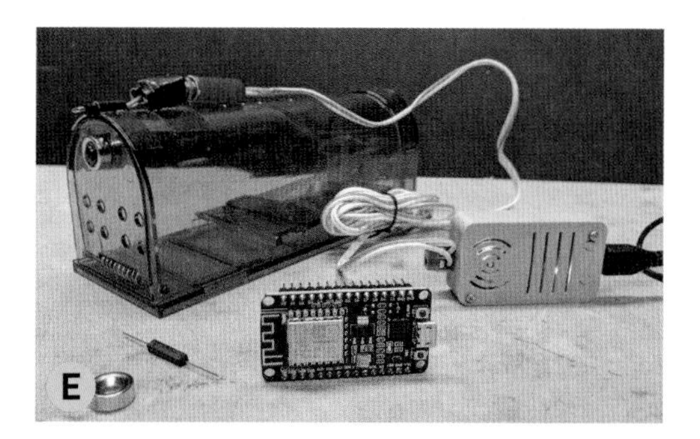

SUMP PUMP ALERT

Like many homes, mine has a sump pump that doesn't run often — but if it's running for more than a few minutes, that's a big concern.

Sump pump pits are not a great environment for electronics. Worse, mine is outside, below grade near a door to the walk-out basement. Not great for radio communications, either.

My solution was to monitor the AC current in the electrical panel for the sump pump circuit (like I did the laundry machines at MIT). There are products available to monitor AC circuits with clamp-on transformers, but it's easy to make one at a low cost. How does this work? A wire carrying an electrical current creates a magnetic field in proportion to the current. A coil of wire wrapped around the hot wire connected to a circuit breaker can measure this current. I used a SCT013 split core current transformer (Figure **F**), readily available via Amazon, eBay, and other electronics vendors.

I put together a circuit to adapt the transformer output current to a voltage in the sensing range of the ESP8266 microcontroller input pin. A Schottky diode protects the input from the voltage spikes produced when the sump pump motor turns on or off. An Arduino sketch reads the voltage and informs Home Assistant.

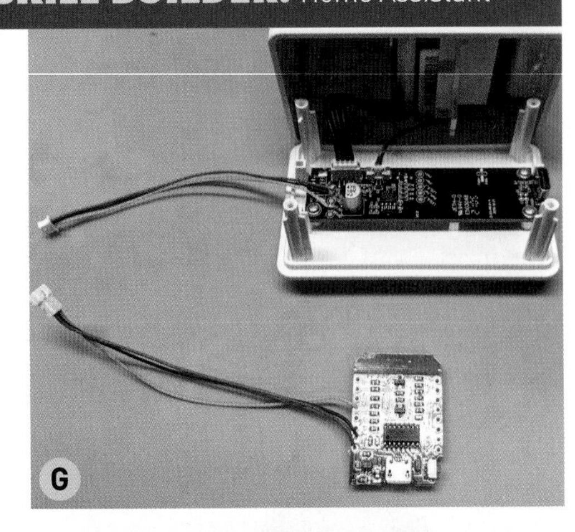

LET'S LINK THE LANDLORD

HOME ASSISTANT IS GREAT FOR RENTAL HOMES By Michael Seltzer

A lot of home automation products get installed into the home physically. This can be a problem or even an impossibility for people who don't (yet) own their homes. So how do renters realize their HA dreams?

I grew up in a connected home (thanks, Dad). Next to my bed was an X10 controller that allowed me to turn my lights on and off without getting up. The unfortunate side effect of this was that my parents unknowingly would turn "all lights off" including mine (maybe it was a hint that it was time to go to bed).

Now I live with my girlfriend and our 1-year-old puppy, and an array of Wi-Fi connected light bulbs, a robot vacuum, and sensors of various types. Tying all this together is Home Assistant, which enables the automation of a motion-activated patio light, a crucial necessity while potty training and letting out a young puppy multiple times a night.

Because I rent, all of this has to be easy to move — no replacing wall switches, for example. Home Assistant allows that automation to be easily built in a world where the motion sensor and the light bulb it controls are not made by the same manufacturer. And I can take it all with me when I move out.

The automation is easy:

- **Trigger:** The sump pump AC current is sensed for more than 1 minute.
- **Action:** Send email and text messages.

INDOOR AIR QUALITY MONITOR

We've seen a lot of wildfires across the USA, and should be concerned about indoor air quality whenever outdoor is impacted. Poor indoor air quality might be addressed by improving HVAC filtration and resolving leaky door and window seals.

On social media I noticed a project that a Home Assistant user had created. This hack adds MQTT to an inexpensive Ikea "Vindriktning" air quality sensor, which happens to have an unused serial port which sends particulate concentration data. Easy to connect this to an ESP8266 serial input pin (Figure G) and use an Arduino sketch to send the data to Home Assistant. Thanks to Sören Beye for sharing his ESP8266 firmware at github.com/Hypfer/esp8266-vindriktning-particle-sensor.

- **Trigger:** The particulate level rises to a specific level.
- **Action:** Send a message, and tell the Nest Thermostat to turn on the fan.

» **Build Your Own IAQ Sensor:** Last summer I led a workshop at the Boulder Public Library for community members to build IAQ sensors and send them to a shared map. Here are the instructions to build your own: bit.ly/aqboulder.

WHAT'S NEXT: MATTER

We asked ChatGPT for a paragraph about **Matter** (csa-iot.org/all-solutions/matter):

Matter is a new home automation standard that aims to make it easier for smart home devices from different manufacturers to work together seamlessly. It was developed by the Connected Home over IP working group [now known as Connected Standards Alliance (CSA)], which is a partnership between Amazon, Apple, Google, and the Zigbee Alliance. Matter is based on Internet Protocol (IP) and uses existing communication standards such as Wi-Fi, Bluetooth, and Zigbee to enable interoperability between devices. This means that a wide variety of smart home devices, including lights, thermostats, door locks, and

Device info

TEST_PRODUCT
by TEST_VENDOR

Firmware: esp-idf: v4.4.3 6407ecb3f8

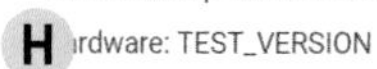

rdware: TEST_VERSION

Controls

ADD TO DASHBOARD

Logbook

January 19, 2023

turned on
9:36:28 PM - 2 minutes ago

turned off triggered by service light.turn_off

more, can all be controlled from a single app or hub using Matter. The goal of Matter is to create a more open and secure smart home ecosystem, and it is expected to be widely adopted by device manufacturers in the near future.

Sure, that seems to be consistent with what we know. But does Matter achieve what CEBus tried to do in home automation standards and interoperability? We hope so!

The January 2023 Consumer Electronics Show (CES) included a plethora of announcements about Matter-compatible home automation devices and products. The CSA certification site lists many Matter products (csa-iot.org/csa-iot_products). We're looking forward to checking some out to see how they improve our current home automation systems.

DOES IT *MATTER* TO MAKERS?

Meanwhile, can we make our own Matter devices? Seems reasonable, given that the Matter standard is published and available to anyone.

In fact, there is a Matter SDK for the Espressif ESP32 microcontroller family (github.com/espressif/esp-matter). Thanks to Jakub Dybczak for developing an Arduino library too (github.com/jakubdybczak/esp32-arduino-matter).

I'm running the "light" example from this library on an ESP-VROOM 32 microcontroller board. The ESP32 includes Bluetooth and Wi-Fi interfaces. In this example, Bluetooth is used to ***commission*** a Matter device, along with a unique ID, perhaps via a bar code. Once commissioned, the device uses Wi-Fi to communicate with other Matter devices.

Using the Home Assistant Matter integration 2023.1.6 release, it works! Home Assistant is able to commission the device (Figure H). Pushing a button on the ESP32 toggles the LED and the status is immediately updated in Home Assistant. Home Assistant is also able to toggle the LED (Figure I). Now that I have a "Hello, World!"

Matter device, I'm ready to make something! What will it be?

Next, Amazon Alexa tried to commission the device. It was detected, accepted a bar code scan, but the Alexa app said "Something went wrong."

OK, how about Google Home? It eagerly reported that a Matter device was available, scanned the barcode, added it to Home, and then the device was nowhere to be found.

None of this is easy to debug at this time. Perhaps the next Matter release (2.0) will address this incompatibility with the ESP32 SDK.

WHAT WILL YOU AUTOMATE?

How can home automation technology make your home smarter? Maybe you already have a project completed. Join the growing community of makers who are sharing their code and hardware implementation with the worldwide community of home automation enthusiasts.

- **Home Assistant Community Forum:** community.home-assistant.io
- **Discord:** discord.com/invite/home-assistant
- **Facebook:** facebook.com/groups/HomeAssistant

WAYNE & MICHAEL SELTZER, father and son home automation (HA) enthusiasts, share their experiences living in homes with HA technology, experimenting, installing, and building their own HA stuff.

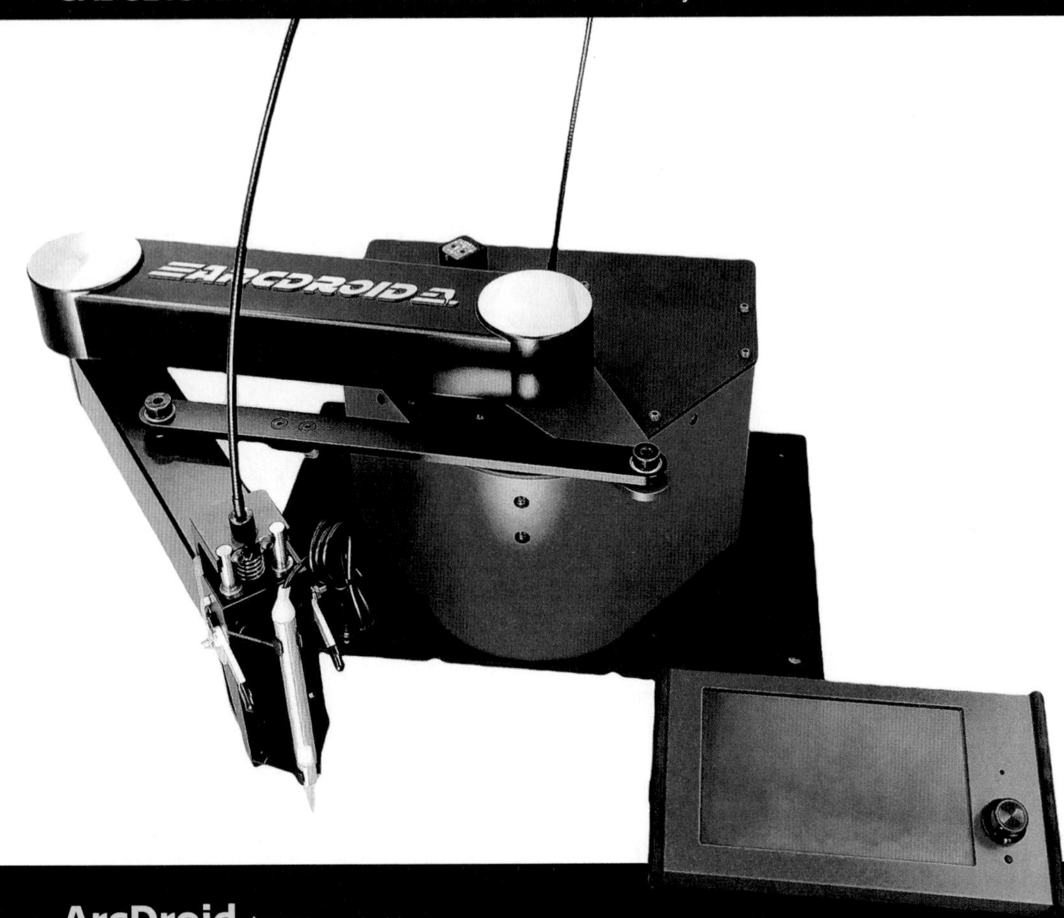

ArcDroid $2,500 arcdroidcnc.com

When it comes to CNC plasma cutters there are a ton of options. Most are 2'×2', 4'×4', 4'×8' and bigger. Prices can range from $2,500 to tens of thousands. I have been looking to get into plasma CNC for a very long time. However, my garage space is very limited, and I've been putting it off for years. I really wanted something that was smaller that I could move out of the way.

I saw an advertisement for a portable plasma CNC called the ArcDroid, and I was instantly intrigued about how it operates. This is unlike any CNC I have ever seen. It isn't your typical gantry CNC plasma table; it uses what's called a SCARA robot arm to pivot on the X and Y axis, which makes it very portable. It can cut a 26"×15" or 660×380mm area, which is perfect for me.

If you are into 3D printers, ArcDroid has some similarities. It runs on a fork of Marlin for its firmware, which is used on a lot of 3D printers. The display is a Bigtreetech 7-inch touchscreen that's also used on 3D printers. Out of the box, it was smaller than I anticipated. Setup was easy: do a calibration, in which you move the arm along a triangle, once with the stylus, and again with the torch. Then you are ready!

I have absolutely zero experience with CNCs and I was able to get my first cuts going in about an hour. I am really happy with my purchase!

Another unique thing about the ArcDroid is that you can trace an object with what they call Simple Trace. There is an included stylus that attaches to the SCARA arm. If you're a cardboard template fabricator, this is right up your alley! You simply trace along the object, put in your parameters and you are cutting! *—John Ivener, Tripod's Garage*

Voxelab Aquila D1 3D Printer

$399 voxelab3dp.com

Voxelab's latest entry into the desktop market, the Aquila D1, isn't intended to make huge headlines. It's more like a nice iteration on the previous Aquila versions. With a roughly 235mm cubed build volume, auto bed leveling, a 300°C capable nozzle, linear X and Y rails, and a removable flexible build plate, it's no slouch.

Assembly was quick and easy. My only complaint is that the cable for the extruder is a little too free and seems to be slightly in the way of the filament path. I guess time will tell if that is a real issue or not.

After running the automatic bed level and beginning the first print, I found I had to adjust the Z-offset to get the hot end down to the bed. This really is no big deal but might confuse an absolute beginner. Once that was done, printing was near perfect. Not only are the prints fantastic quality, the bed is excellent as well. Though it is removable, the prints have been popping off on their own and I haven't even needed to flex it. The textured bed leaves a good feel to the bottom of your prints too, which is nice.

At the price this is offered, it delivers fantastic print quality, and that's the bottom line. I'd happily have this printer in my collection based on the quality of the results. *—Caleb Kraft*

Cuttle.xyz

Free (for personal use) cuttle.xyz

Cuttle is an online editor for creating files for laser cutting and CNC cutting. It is surprisingly powerful, even in the areas that are completely free for personal use. There's a general drawing tool, but also some stand-alone templates that allow you to create things like boxes, keychains, or envelopes, with sliders to adjust parameters like size, all exported with the optimal file for laser cutting.

There seems to be a pretty big gap in design software for vector files for this kind of thing, and the folks at Cuttle have really put some thought into what is needed for proper lasering. The paid features look pretty useful too.

In addition to the free tools and paid tools, there are a few very well done tutorials. *—Caleb Kraft*

TOOLBOX

GADGETS AND GEAR FOR MAKERS

Tell us about your faves: editor@makezine.com

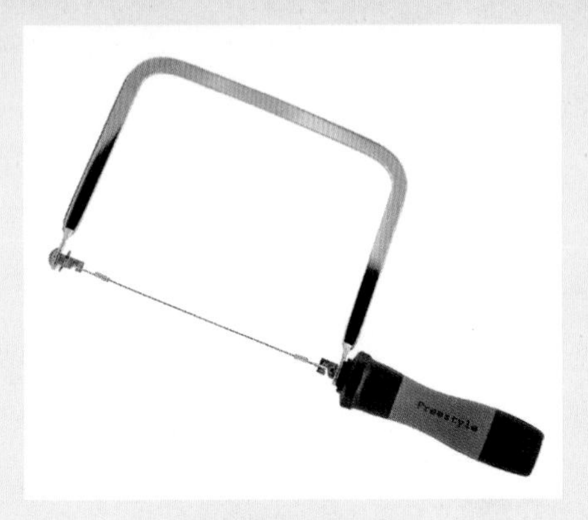

Spyral Freestyle Coping Saw & Spyral Saw Blades

Saw $18, Blades $6–$8 for multi pack
spyralsaw.com

Cutting curves in thin sheet stock with a coping saw is major pain — attempt too tight a curve, or let your stroke stray from perfectly perpendicular, and the flat blade abruptly sticks and binds, bending your workpiece and causing tearouts or cracks (and often a horrendous screech).

Instead, try these wire blades with a spiral tooth, from Bestway Products. They're like those wire survival saws (Chuck Yeager cut off his helmet with one after a crash, badass!) but made for coping saws, bandsaws, hacksaws, and jeweler's/scroll saws too. Turn any corner you want, this blade does not want to bind. Using their Spyral Freestyle coping saw, I easily cut tight little ¼" keyhole shapes in floppy ⅛" acrylic sheet, swinging the saw around to apply pressure from any direction that felt right. Smart. *—Keith Hammond*

Bollé Safety Glasses

$9 and up bolle-safety.com

I thought Bollé just made glacier glasses and other spendy shades for outdoor sports. So I was stoked to find their stylish, affordable line of indoor and outdoor safety glasses, from basic industrial ($9 and up, $15 and up for polarized) to mil-spec ballistic ($20 and up), to blue-light ($74 and up), in standard and Asian fit. Pay a little more for anti-fog and specialty coatings. Now whether I'm stirring resin, drilling out stripped screws, or sanding in the sun, I have eye pro that's not an eyesore.

—Keith Hammond

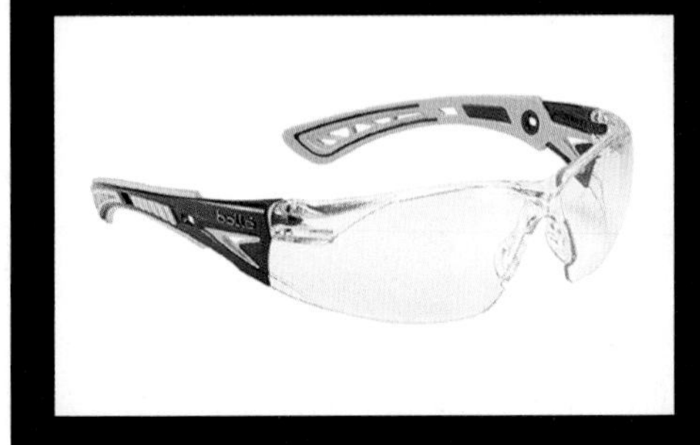

ILLUMINATING THE DEEP

Susan Ebberson

It takes a village to raise a child, and it takes a team to fabricate a 20-foot techno-cephalopod. With a glistening skin composed of colorful ceramic tiles, and LED-lit plastic domes for suckers, *Lucy* the octopus is the brainchild of artist Peter Hazel, who has made a name for himself creating giant animals such as the massive crocodile *Niloticus* and 10-foot-tall rattlesnake *Green Mojave*.

Hazel and his team of metal fabricators, ceramicists, and welders have been hard at work bringing this eight-armed piece of art to life. Keith Williamson of Electric Fire Design has been toiling to bring light into the 395 luminous accents, coming up with custom electronics and code. The LEDs flicker and fade to accentuate the flowing lines of the sculpture, and the light patterns and designs can be changed on location by a non-programmer without any need to recompile software.

Lucy's final home is slated to be a very public space in California where she will be a delight for all

US $14.99 CAN $17.99
ISBN: 978-1-68045-800-8